BUILDING CARE

BUILDING CARE

Brian Wood

Field Chair – Building Care and Head of Quality Assurance
Centre for Construction Management
Oxford Brookes University

Blackwell
Science

Editorial Offices:
9600 Garsington Road, Oxford OX4 2DQ, UK
Tel: +44 (0)1865 776868
Blackwell Publishing, Inc., 350 Main Street, Malden, MA 02148-5018, USA
Tel: +1 781 388 8250
Iowa State Press, a Blackwell Publishing Company, 2121 State Avenue, Ames, Iowa 50014-8300, USA
Tel: +1 515 292 0140
Blackwell Publishing Asia Pty Ltd, 550 Swanston Street, Carlton South, Victoria 3053, Australia
Tel: +61 (0)3 9347 0300
Blackwell Verlag, Kurfürstendamm 57, 10707 Berlin, Germany
Tel: +49 (0)30 32 79 060

First published 2003 by Blackwell Science Ltd

Library of Congress
Cataloging-in-Publication Data
Wood, Brian.
Building care/Brian Wood.
p. cm.
Includes bibliographical references and index.
ISBN 0-632-06049-2 (Paperback)
1. Building–Maintenance. 2. Buildings–Repair and reconstruction. 3. Architecture–Conservation and restoration. I. Title.

TH3351 .W66 2003
690′.24–dc21 2002154157

0-632-06049-2

A catalogue record for this title is available from the British Library

Set in 10/12.5pt Palatino
by DP Photosetting, Aylesbury, Bucks
Printed and bound in Great Britain by
TJ International, Padstow, Cornwall

For further information on
Blackwell Publishing, visit our website:
www.blackwellpublishing.com

Contents

Foreword

Maintenance work generates nearly half of the output of the UK construction industry. It is absolutely vital for everyone in all walks of life, be they people in their homes, workers in offices, factories or shops, teachers and pupils in schools, doctors, nurses and patients in hospitals, or myriad of others. Yet public discussion in the industry tends to focus on glamorous new construction, which actually touches the daily lives of most people far less. No-one wants to live in a cold or leaky building. Yet maintenance remains a Cinderella of the construction process, little discussed and often ignored in industry and media circles, except when vicious cowboys rip off some elderly householder.

Brian Wood rightly confronts that dismissive attitude throughout his thoughtful and profound analysis of the maintenance process. He begins with the link which is often missing in construction debate – the client comes first. It is not just that the client's wishes ought to be met by the supply side. That should be automatic. It is that the client should always be at the core of the process. That must begin with the initial design of any new or refurbished building, with the client's facilities or estates manager being fully involved. This basic requirement of client satisfaction is well dealt with by the author, who rightly stresses that new procurement routes and IT capabilities make such effective involvement of the client a practical reality. Many expert clients now insist upon it, and the message needs to be rippled down to lay and occasional clients.

The author also tackles the industry's biggest 'cop-out', that 'it is different, and has nothing to learn from other industries'. That is what the British car industry also believed in the 1950s. The Japanese taught it a harsh lesson, but not until the British car-makers had tested their own negative and self-absorbed theory virtually to destruction. 'Just in time' techniques are equally relevant to building maintenance as they are to assembling a pop star's Ferrari or a family hatchback. They are about disciplined work flow.

So also is effective re-engineering of that process. The building industry needs to accept that there must be a process in the first place, which many still do not, and that the process is there to deliver client satisfaction. It is not just about 'after care' at call centres. Rather, it is about effective initial design to produce intelligent buildings, sustainable techniques, modern value management and flow processes which involve the client at every stage. All that is expected from an industry which does not greatly value maintenance. It also requires new thinking from the general public, where the assumption is still that 'maintenance' is the bloke with a spanner and dirty overalls who does not come on time or do a good job, rather than highly skilled professionals, business techniques and full committment to service delivery at every stage.

Putting maintenance at the centre of the continuing post-Egan reform process is long overdue. There is much more to the industry than great structures such as a bridge or a PFI hospital. New or re-furbished projects also need to work properly and be in continual use, not moth-balled because of defects. This book seeks to advance that debate and the reform process in a very focused way. I warmly commend it to you and to a wide audience.

Sir Michael Latham
DL, MA, (Hon)FRIBA, FICE, FASI, FCIOB
Author, *Constructing the Team*

Preface

Building care encompasses the management and maintenance of premises. Premises-related costs typically represent 15% of the costs of running a business; they are the second largest cost after people and they are the largest area of readily reducible costs to seriously impact on the bottom line. While facilities management is still a relatively young and exciting area of management, building maintenance, though expensive, is lowly regarded; it is not sexy.

Building care is customer focused. Previous paradigms such as planned and so-called preventive maintenance (PPM) and the rationale behind them have been rendered possibly redundant by developments in technology and management techniques. The rationale and conditions that supported PPM have been overtaken by changed commercial and environmental imperatives that are driving new approaches. A range of new, user-friendly approaches are discussed and evaluated.

This book offers help and guidance to those responsible, from board level down, for commissioning and carrying out efficient and effective building care services. It is essential reading for professional facilities and estates managers, company directors, architects, engineers, surveyors and maintenance managers. It is also relevant to students of those and related disciplines at technician, undergraduate and postgraduate levels. It will also guide the reader to further sources of edification and enlightenment.

Brian Wood

Acknowledgements

Firstly, I thank my colleagues and anonymous industrial collaborators at the International Council for Research and Innovation in Building and Construction and throughout continental Europe, North America, Africa, South-East Asia and Australasia for enhancing the scope and scale of my studies through friendly and convivial advice and guidance; and 'students' young and not so young, who have questioned everything.

I am deeply indebted to my wife, Erica, for her encouragement to bring together these developing strands of thought, and her forbearance and endurance in then putting up with my highs and lows of producing this book. I am also grateful to those who gave me opportunities to develop ideas and share experience and to the many colleagues who have encouraged me along the way.

From my first employment as an architectural student at the London Borough of Hillingdon I learnt the value of building on tried-and-tested details while injecting something new to each scheme. Following Thurston Williams to the National Building Agency, I got used to 'thinking big' as we developed ten-year rolling programmes of major repairs and improvements for housing and other building stocks around the UK. David Bland and Frank Bradbeer taught me much about teamwork and long hours.

I acknowledge the prompting of Trevor Eastell of Richard Hemingway and Partners and the late Roy Holmes in encouraging me into teaching, and the help of many more in developing my research interests. Particularly, I thank Hedley Smyth for joining with me in the research which gave birth to 'just-in-time maintenance', and Chris Topley for supporting the project.

Finally, it is never possible to give due credit to all those who played a part in bringing a project to successful completion, so for all those I forget to mention and those to whom I give inadequate credit, I hope I may be forgiven these sins of omission. For all other errors, especially where I have effected to present my opinions as facts, unsupported by evidence, I take full credit.

1 Background and Introduction

The last decade of the 20th century and the start of the 21st have seen significant growth and development in the discipline of facilities management. This has been in part a response to a changing business environment in which organisations have sought to return to core business or privatise operations. Processes have been re-engineered and customer focused as facilities management functions have been subcontracted or outsourced and provided remotely. Developments in technology and management techniques and priorities have put increasing power of various kinds in the hands of multifarious and diverse building users, and have facilitated new approaches to the maintenance and management of premises. Bringing these innovations together under the flag of Building Care reflects a renewed interest in these vital yet undervalued activities, and encourages further thought on how professional premises-related services may best be provided to meet the changing needs of today.

This introductory chapter examines the context in which maintenance takes place, where often buildings are neglected over long periods until something must be done to repair, rectify or improve the situation. Subsequent chapters consider ways in which those responsible for maintenance in a professional capacity have been trying to develop policies and processes to address current deficiencies, before moving on to conceptualise alternative approaches to achieving buildings that may be more appropriate for the future.

Cinderella and the cowboys

The maintenance of property and estates has long been portrayed as unattractive. Seeley (1976) described building maintenance as something 'regarded by many as a 'Cinderella' activity (with) little glamour, unlikely to attract very much attention and frequently regarded as unproductive'. In this he reflected a 1972 UK government committee report on building maintenance that described how 'this class of work is accorded little or no merit'. Milne (1985) talked of maintenance as 'a slightly inferior branch of the industry'. Michel, in his foreword to the CIOB *Handbook of Facilities Management* (Spedding, 1994), states 'for far too long the management and maintenance of the built environment has been the poor relation of the new-build sector of the construction industry'.

Seeley calculated that 'the output of the construction industry on maintenance work in 1969 represented 28% of the total construction output for that year'. However, a more recent report (*Building Design*, 1998) declared 'A booming building maintenance and refurbishment market could be worth almost three times the new-build market'. The study on which that report was based (Barbour, 1998) estimated that the UK market for maintenance and related work was £28 billion (45 billion euros) compared with £10 billion (16 billion euros) for new buildings.

Cowboys

Despite the size of the building maintenance market and the value of built assets, much external perception of 'building' and 'maintenance' is of 'builder's bum' and 'cowboys'. 'Cowboy' builders are generally unqualified and unprincipled individuals undertaking building work. Often, their work will be of dubious quality, and this contributes to a poor public image of builders and the construction industry generally. In the UK, there is no requirement for a builder to be qualified nor to be in any way registered. Steps are being taken to redress this situation. For instance, the UK Department of the Environment, Transport and the Regions set up a working group under Tony Merricks, MD of Balfour Beatty, a global player in the construction industry, to take forward an initiative 'Combating Cowboy Builders' (DETR, 1998a). The executive summary of a consultation document (www.construction.detr.gov.uk/consult/cowboy) identified that 'tackling the problems caused by so-called "cowboy builders" in the repair and maintenance sector will require action on a broad front to improve service to the customer and raise standards'.

A detailed overview of the DETR proposals and responses to the consultation by contractor organisations and professional institutions was presented at a research conference organised by the Royal Institution of Chartered Surveyors and subsequently published (Wood, 1998). That paper also identified the nature of contractors serving the UK building maintenance market, indicating, for instance (Rolfe & Leather, 1995), that:

- most firms ... are individuals or firms with less than six staff
- only half have a formal construction industry qualification
- few of the contractors have formal training in running a business
- only 14% ... employed administrative or secretarial help.

The poor shape and structure of the building maintenance industry is discussed in more detail in Chapter 4. In the UK this manifestly unsatisfactory state of affairs is being challenged by the entry of 'non-traditional' players. A paper by Cahill & Kirkman (1994) reported that National Breakdown (now Green Flag Home Assistance Services) had 'used the knowledge and skills developed for the motoring organisation to expand into home emergency services'. This service, referred to as 'call-centred maintenance' (Wood, 1998), is also described more fully in Chapter 6.

In Australia, a similar concern about poor building services has been addressed through legislation. For instance, under the State of Victoria Building Act 1993 builders who undertake domestic or commercial building work are required to register as building practitioners. Qualifications and experience required for registration are prescribed in the Building Regulations, and appropriate insurances must be in place. A useful study of the effect of this legislation has been published by Georgiou *et al.* (2000). Although 'too early to state categorically' it seemed that the benefits outweighed drawbacks, with consumers appreciating the protection afforded.

Property conditions

Not only is much of the construction industry in poor shape to provide good maintenance services, the buildings to be maintained are often in poor condition. In the UK and elsewhere, it has been a common practice for many building types, for instance churches, educational and health care institutions and other publicly owned buildings, to carry out periodic reviews of building conditions, perhaps a 'quinquennial inspection'. Typically, these would be undertaken by an architect or surveyor and give rise to a schedule of works to be carried out over a number of years and prioritised according to criteria which may or may not be explicit. There may be measures of scale as well as time; there may be repair/replacement/improvement recommendations and there may be indicative specifications and/or costs. The following are examples.

A study by the Architects and Building Group of the former Department of Education and Science (1985), reported by Shen (1997), identified the following three-point scale as 'typical'.

- *Priority 1* – work needed immediately or in the near future to meet legislative or contractual requirements and to ensure the health and safety of building occupants and users; work required to prevent the imminent closure of accommodation or serious dislocation of activities.
- *Priority 2* – work necessary within one year to prevent serious deterioration.
- *Priority 3* – work as above which may be deferred beyond one year; work desirable to maintain ... environmental quality.

In the UK National Health Service there has been a consistent use of a simple alphabetical code to describe condition by reference to that achieved by a new building or component (NHS Estates, 1995).

- *Condition A* – the element is as new and can be expected to perform adequately for its normal life.
- *Condition B* – the element is sound, operationally safe and exhibits only minor deterioration.
- *Condition C* – the element is operational but major repair or replacement will be needed soon, within, say, three years for building and one year for an engineering element.
- *Condition D* – the element runs a serious risk of imminent breakdown.

Pitt (1997) offers an example of a chronological categorisation.

- *Condition 1* – where the component or element needs replacement or major repair within one year.
- *Condition 2* – where the component or element needs replacement or major repair after one year but before three years.
- *Condition 3* – . . . after three years but before six years.
- *Condition 4* – . . . after six years but before ten years.
- *Condition 5* – . . . does not need replacement or major repair for ten years.

A similar, though six-point, time- or lifetime-related scale has been used in Holland (Damen *et al.*, 1998).

(1) New building condition.
(2) None or only slight defects, visible symptoms of ageing.
(3) Many slight and/or some substantial defects, over half a lifespan elapsed.
(4) Substantial defects, remaining lifespan limited and not difficult to predict.
(5) Very substantial defects; component has reached end of its lifespan, replacement or major repair necessary.
(6) Component is completely worn out; almost total performance loss; should have been replaced earlier.

A matrix combining scale and timescale was used by the UK National Building Agency:

2	Repair	Early
3	Repair	Middle
4	Repair	Late
5	Replace	Early
6	Replace	Middle
7	Replace	Late

Code 0 was used where no work was required; code 1 indicated that a note was attached; code 8 for 'further investigation required'; and code 9 for an emergency situation. 'Repair' indicated any work less than a total 'replace'; 'early' would be within the first one or two years of a forward programme of work; 'middle' and 'late' accordingly within a projected five- or ten-year horizon.

English House Condition Survey (EHCS)

If similar surveys are carried out at regular intervals it will be possible to plot changes in condition, whether of decay or improvement, over time and thereby to inform action. For instance, in the UK since 1971 there have been five-yearly surveys of house conditions in the constituent parts of the kingdom. A longitudinal study of changing house conditions has been possible by the survey of a constant sample of buildings using a constant survey pro forma and consistent criteria for assessment of condition, in theory. More will be said in Chapter 2 about difficulties in practice revolving around developing pro formas (from a one-page form in 1971 to four pages in 1981 and 14 in 1986), surveyor variability and changing standards over time. A good summary of the English House Condition Survey and its development has been produced by Davidson (1995).

The first EHCS was carried out in 1967 to provide the government with more information to inform policy development regarding older housing. That survey 'showed that there were twice as many unfit dwellings as had been believed' [and consequently] 'the 1969 Housing Act increased the availability of grants for owners to improve their properties...' (Davidson, 1995).

Reports analysing changes in condition between one survey and the next have fairly consistently indicated that despite repair and improvement interventions on the housing stock, overall conditions have improved little. In essence, repair works just about keep up with incipient decay. There is a case for suggesting that it may always be thus; that either there is a consistent threshold of condition acceptability that 'triggers' a repair or replacement of a component, or that perhaps surveyors' assessments of condition are 'conditioned' by reference to increasing standards over time.

Repair backlogs

A report by the Confederation of British Industry in 1985 presented 'evidence of the backlog of essential repair and maintenance work ...

- As long ago as 1977, a Department of Education and Science report, 'A Study of School Buildings,' indicated that £1.2 billion (£2.5 billion at 1985 prices) needed to be spent to bring schools built before 1976 up to modern standards.
- In 1983 the Davies Report 'Underused and Surplus Property in the National Health Service' put a figure of £2 billion on the amount that should be spent to bring hospitals and health centres up to minimum acceptable standards.
- An Audit Commission report this year [i.e. 1985], 'Capital Expenditure Controls in Local Government in England', stated that the routine maintenance backlog for council housing has been accumulating at the rate of some £1 billion per annum "for a number of years".'

In the same year (1985) the then UK Department of the Environment published a study of the condition of local authority-owned housing stock. A review of that study (Wood, 1986b) indicated that an estimated £19 billion of expenditure was required to bring the properties to an appropriate condition. By comparison, a report on conditions some 15 years later referred to 'Deputy Prime Minister John Prescott's surprise estimate last week of a £19 billion council repairs backlog' (*Inside Housing*, 14 April 2000).

The English House Condition Survey carried out in 1996 (DETR, 1998b) showed continuing disrepair and unfitness in the housing stock.

- Nearly 80% of dwellings had some fault recorded to the external or internal fabric.
- The average level of disrepair across the whole stock was about £1500 for an average-sized house. The mean costs of repairs and replacements due over the next ten years for owner-occupied semis and terraces built pre-1919 were £3460 and for private rented pre-1919 houses were £5320.
- There were over 1.5 million (1,522,000) unfit dwellings, representing 7.5% of the stock. About 1 million of these dwellings were also recorded as unfit in 1991 while the remaining half a million became unfit after 1991.

This evidence suggests that while there is investment in the improvement of existing housing stock, it is just about keeping pace with decay; so, while some properties move out of unfitness or disrepair, others move in. With a profile of increasingly ageing stock this may mean an increasing need for expenditure to stave off disrepair. Correspondingly, an increase in disrepair may be expected if expenditure is not increased.

There is also evidence that suggests that some of the newer housing is of less robust construction than some of the older stock; this may require even higher expenditure. Perhaps, like the poor, repair 'backlogs' will be always with us.

Studies of other building types such as schools and health service buildings have also shown substantial backlogs of maintenance. Shen (1997) noted that 'The report from the former Department of Education and Science (1985) also showed that backlog of main structural works in primary schools and secondary schools approximate to £490 million and £370 million, respectively'. He also quoted the Society of Chief Architects in Local Authorities (SCALA, 1993): 'the average actual expenditure on repairs and maintenance of educational buildings was £8.04/m^2 which was far lower than what was required, £20.25/m^2, to maintain the buildings in a safe, dry, warm and comfortable condition'.

Condition or performance-based maintenance

Maintenance, as defined by British Standards BS 3811 and BS 4478, is work intended to restore a service to a satisfactory working condition. It cannot reasonably be expected to restore service to a fully 'as new' condition. How is one to define the 'acceptable' condition? It is possible to picture the life of a building or component as having a level of performance that declines over time, with maintenance interventions at intervals (which may or may not be regular intervals) intended to restore a higher level of performance. That level of performance may be higher or lower than the initial performance level, depending on the nature of the repair, replacement or improvement.

The point in time at which a maintenance intervention is made may be determined in relation to a defined lowest acceptable performance level (LAPeL). If that is the point at which the maintenance action is initiated then there will be a period of time during which the component is performing at an 'unacceptable' level. This may be regarded as 'acceptable' if it is for only a defined period of time, perhaps defined by a service level agreement.

Alternatively, a maintenance intervention may be 'triggered' at a point when performance of the component in question is sufficiently above the LAPeL that the maintenance action is complete before performance would have declined below the LAPeL. This point may be determined by projections of performance based on records and/or monitoring. In some circumstances this could be informed by a decision support system and/or could be implemented through automation. (Decision support systems were previously known as 'expert systems'.) The closer the 'trigger' point is to the LAPeL, the greater the effective utilisation of the component. Correspondingly, the closer these two points, the greater the risk that the LAPeL will be breached. In some circumstances this may be operationally unacceptable, perhaps infringing safety or statutory requirements. By contrast, if the intervention point is significantly above the LAPeL, utilisation may be unnecessarily curtailed. Issues related to automatic or automated maintenance actions are discussed in more detail in Chapter 7.

Repair or replace

The recognition of repair 'backlogs' demonstrates that maintenance is not always carried out when someone thinks it should be. Buildings are robust. For instance, a component will still provide service beyond the point at which someone believes it should be repaired or replaced. It is a matter of judgement. There are many considerations to be made in determining the most appropriate timing and scale of maintenance action, including, for instance:

- costs, including changing costs over time and costs in use
- availability of finance
- downtime
- effects of service failure
- convenience
- availability of parts/spares
- changing standards
- staff/skills
- acceptable performance
- warranties/guarantees/service agreements
- access
- value(s).

Maintenance engineers or surveyors, or whoever is responsible for making such repair/replacement decisions, will be weighing many factors in the balance and people will reach different conclusions in different circumstances.

Decision making also consumes professional staff time and therefore also costs. Is it better to make decisions quickly or to make the right decisions? It is useful to have mechanisms for assisting expeditious decision making. Such protocols may be assembled in maintenance manuals or incorporated into computerised systems. As Then (1995) indicates, 'It is hoped that the incorporation of condition surveys will bring about a proper balance in deciding on priorities between the physical needs of the stock and demands of political expediencies and pressure groups'. Often repair and replacement works will be assembled into composite programmes of work, perhaps a planned preventive maintenance (PPM) programme. The provenance and pitfalls of such programmes are discussed in some detail in Chapter 2.

Deferred maintenance

'The best laid schemes o' mice an' men gang aft a-gley' (Burns, 1786).

In essence, our schemes, however well worked out, often do not go quite according to plan; revision is generally required, sometimes major revision. A common problem is that insufficient funds are made available. Maintenance is a frequent victim of budget cuts or pruning of programmes; it is more usually seen as discretionary spending rather than as investment. There is a strong argument that expenditure deferred is expenditure saved.

Counter to the above argument is that 'a stitch in time saves nine'. The suggestion here is that deterioration proceeds over time such that whereas at one point in time a minor repair, maybe a mere redecoration, may suffice, at a later time a proportionately more major repair or replacement may be required. 'Spoiling the ship for a halfpennyworth of tar' is another expression in related vein, suggesting the substitution of the required repair with a lesser action that may save money in the short term while deferring greater expenditure to a later date. Deferred maintenance may equate with deferred expenditure, but that is far from the same thing as expenditure saved.

What is implied by deferred maintenance is an acceptance, whether explicit or implicit, of a reduction in performance or perhaps a substitution of a planned maintenance action by a maintenance intervention of an emergency kind at an uncertain and maybe inconvenient time. Thus deferred maintenance, while it may be planned, is the antithesis of planned maintenance, in that maintenance will not take place at the otherwise optimal time. It is also less likely to be preventive as there will be an increasing risk of service failure during the period of deferral.

Influence of the public sector

As economies and social priorities around the world grow and change, so do the size and involvement of the public sector. For instance, in the decades immediately following the two world wars, there were significant building programmes in which governments were major participants. For example, in the UK there were extensive programmes for councils to build 'homes fit for heroes' in the 1920s and 1930s. Many a provincial town has its Addison Crescent and Asquith Road built at that time. Addison was the Minister of Health in Asquith's government which in 1919 'decided to open the entrepreneurial coffers of the local authorities' (Pawley, 1971), heralding the birth of 'council housing'.

The Addison Act was succeeded in 1923 by the Chamberlain Act, which offered financial aid to private builders and local authority mortgages for those with sufficient savings to purchase their own dwellings, thus swinging public sector support behind private sector housebuilding. Similarly, after the Second World War there was a substantial programme of council housing construction superseded in the 1950s by Macmillan's 'never had it so good' era of mass private housebuilding, extending Metroland and similar suburban developments. 'Let us be frank about it: most of our people have never had it so good' (Macmillan, 1957). In the 1960s the public sector came back to centre stage with tower blocks and prefabricated 'large panel system (LPS)' building in the ascendancy until the gas explosion in Ronan Point shattered confidence in large-scale public housing and redevelopment generally. This sparked a renewed interest in the retention and rehabilitation of older housing which a few years earlier would have been scheduled for 'slum clearance'.

In the 1970s, planning and project management approaches and techniques developed in the public sector for housing construction by the thousands were transferred almost seamlessly through organisations like the UK's then National Building Agency to massive programmes of housing rehabilitation. 'Housing rehab' schemes, following the model of the Ministry of Housing and Local Government's Deeplish Study (MoHLG, 1966), were developed for pre-First World War housing across the country. Repair and improvement priorities were determined by surveys using the English House Condition Survey methodology. This was coupled with legislation which over subsequent years has allowed for the designation of Comprehensive Development Areas, Improvement Areas, General Improvement Areas, Housing Action Areas and, most recently, Neighbourhood Renewal Areas and accompanied by the availability of financial assistance to building owners. With each new government came a new Act and new, usually more restrictive financial criteria.

It is this heritage of public sector planning and intervention in housing, and also schools and hospitals, that has influenced much of the theory and practice of building maintenance until recently.

Programmes

A key component of the 'public sector' approach to building maintenance is the 'programme'. Development of the programme is an important project and so too is its implementation and its management. Typically, a programme will involve the collection and analysis of much data and the determination and application of multiple criteria in weighing priorities. Preparation of a programme is a professional activity requiring many person-days of effort. For instance, the production of a five-year rolling programme of elemental repairs and improvements for a typical UK local authority or housing association stock of around 5000 dwellings may be expected to take around 4–500 person-days utilising a team of 2–4 professional surveyors over a period of 8–12 months. Duration would also depend on the geographical and age distributions of the stock, survey and output specifications, etc. This is a substantial, and recurrent, commitment of time and money; however, it may represent only a fraction of 1% of the asset or rebuilding value of the properties.

The Royal Institution of Chartered Surveyors has recommended that building owners budget around 1.5–2% of the asset value for expenditure on the maintenance of a building. Clearly, some building types and constructions will be more 'hungry' than others and this will also vary over time. The figures are useful for ballpark estimating and may provide a basis for assessing the total value of a programme taken over a span of several years. Deviations from norms must be expected and should be carefully evaluated. The development of performance indicators (or key performance indicators – KPIs) is enabling 'benchmarking' studies to be undertaken in many sectors; these are discussed in more detail in Chapter 10.

A study by the Chartered Institute of Public Finance and Accountancy (CIPFA, 1983), analysed by the author (Wood, 1986a), identified wide variations in expenditure. In the year under review (1982–3) local authority-owned dwellings in Kingston-upon-Thames had an average £366 spent on their maintenance, whereas an average £140 was spent on local authority-owned dwellings in Kingston-upon-Hull. Why? Kingston-upon-Thames is an attractive suburban area on the south-western edge of London, built up largely in the mid-to-late 20th century and very much 'semi-detached'. By contrast, Kingston-upon-Hull, more usually known as Hull, is a city built on fish landed fresh from the North Sea, located on the coast of north-east England and with a historic core of older housing. Perhaps the council houses of Hull had been better maintained than those in Kingston in the years immediately preceding and therefore were less in need of maintenance.

Perhaps the council houses of Hull were more modern or better built than those in Kingston? Perhaps council houses in Hull were undermaintained, unloved and decaying; those in Kingston well maintained or overmaintained? Maintenance personnel paid more in Kingston? Maintenance staff more efficient in Hull? A good set of questions, all worthy of being addressed; no simple answer. What would be an appropriate figure for Kingston, Jamaica?

A programme must also be expected to vary over time. Because of interdependence of work elements, it is often not easy to make changes. The deferment of maintenance due to cuts in funding has already been referred to; this may have consequential effects. For instance, it does not make operational sense to renew a pitched roof covering of tiles or slates and shortly afterwards to carry out a deferred rebuilding of the related chimney stack. Not only will there be two lots of scaffolding costs but roof materials may be broken and flashing details will be more difficult to execute with certainty of waterproofing. Furthermore, there will be greater overhead costs and more disturbance of occupants. There is therefore an in-built inflexibility that accompanies a programme which militates against the ability to accommodate to changing circumstances commonly expected of today's businesses.

Direct labour organisations

A further component part of the 'public sector' approach to maintenance has been the execution of work by directly employed maintenance personnel. In the UK the efficiency of this arrangement came under close scrutiny during the governments of Margaret Thatcher. Her antipathy towards the public sector in general, and organised labour and the trades unions in particular, has been well enough documented elsewhere. Suffice it to say here that legislation was progressively introduced that required local authorities' own maintenance departments to tender for work in competition with private building contractors. The underlying rationale was that cosseted council workers and practices protected from the rigours of the market were inefficient and possibly even corrupt. There were possibilities inherent in the 'keeping busy' of a standing labour force that otherwise unnecessary or non-urgent work was undertaken or that staff in some trades may be underemployed at times if there were a mismatch between labour 'supply' and lack of work 'demand'.

It could therefore be the case that work programmes may be biased towards certain trades, in proportion to staffing levels, rather than strictly in relation to condition of the element. In the political climate of the 1980s it was generally unlikely that a programme would make much reference to the needs of building occupants, other than perhaps a 'consultation' with tenants as represented by their elected councillor or notification to a head teacher of work to be carried out that year.

The political climate at central and local government levels has changed much over recent years. There is now a much greater expectation of service and recognition that councils and governments are indeed intended to serve their customers. This will be discussed further in Chapter 6.

Planned economy

In some ways the preparation and execution of comprehensive programmes of work, especially by 'the council', may be considered a vestigial hangover from the planned economy, or 'nanny state' as it has been pejoratively represented. An underpinning logic that 'mother knows best' has been largely superseded by an expectation of 'getting what I am paying you for'. This holds good not only at home, where increasingly people in the UK either own their own homes or expect to be treated as tenants in the same way as if they did own the dwelling which is their home, but also at work. In the workplace again people have increasing expectation of being 'in control' of their working environment. At play too, in leisure and entertainment buildings or shopping, customers are choosing to spend their time or pound or euro or dollar in facilities that reflect their values. People with choice are no longer prepared to patronise premises that are down-at-heel and dowdy. A bright, well-maintained environment, serviced by attentive and well-trained staff, effectively 'sells'.

Privatisation

Policies pursued by Margaret Thatcher in the UK, by Ronald Reagan and George Bush in the USA and by like-minded politicians around the world effectively 'privatised' much of what had previously been considered the domain or prerogative of the public sector or 'the state'. 'There is no such thing as Society. There are individual men and women, and there are families' (Thatcher, 1987). In the UK the period of Conservative government since 1979 saw the denationalisation and privatisation of the air, bus, coal, electricity, gas, rail, steel, telephone and water industries and consequently the management and maintenance of their often extensive portfolios of land and property holdings. For some, this was too much: 'First of all the Georgian silver goes, and then all that nice furniture that used to be in the saloon. Then the Canalettos go' (Macmillan, 1985).

Many 'businesses' previously in some kind of public ownership or management have been given a more 'arm's length' relationship, although they may still be reliant upon government for much of their funding. For instance, the UK's 'new universities' receive funding through government agencies such as the Higher Education Funding Council for England (or their equivalents in the other parts of the country) and the various research councils. They are then responsible for their own management, including the management and maintenance of their buildings and estates. Previously, as polytechnics, they were managed by their respective county councils or local education authorities (LEAs).

Thus many 'private' organisations have come into being from 'public' origins, with 'inherited' staff and inherited procedures. Hence a public sector mentality and associated processes, characterised as 'bureaucracy', 'red tape' and 'restrictive practices', are often held to prevail. A 'culture change' with more entrepreneurial attitudes and outlook may take time to pervade, especially as people are generally suspicious of, and resistant to, change. More about 'change' is discussed in Chapter 9.

It is interesting to introduce here a note regarding housing sector reform in Russia.

> 'The World Bank today [7 May 1996] approved a $300 million loan to help Russian enterprises focus on restructuring their core business activities by divesting their responsibility for housing. The project [total value $551

million] will support reforms designed to push the ownership, management and financing of enterprise housing stock toward the private sector.' (World Bank, 1996).

In 1992, 'enterprises' (previously state-owned businesses) owned 40% of the urban housing stock, city governments slightly less and private owners and co-operatives almost 20%. From 1992 to 1994 there was a programme of mass transfer from enterprises to local authorities. Since enterprises neither owned the housing stock nor employed the tenants, they had little incentive to maintain these stocks. As a result, housing conditions had deteriorated rapidly. Local authorities were concerned that this would 'overwhelm their poorly managed municipal maintenance units'. This led to the proposed 'restructuring' plan, which includes:

- mass privatisation of the housing stock
- targeted housing allowances to protect needy groups
- competitive bidding for housing maintenance, improving quality
- improved energy efficiency, also reducing maintenance costs and improving affordability
- urgent repairs and rehabilitation
- project management by the Central Project Implementation Unit in Moscow and local implementation groups.

Home ownership and 'right to buy'

'For a man's house is his castle' (Coke, 1628). This epitomises the English aspiration to own one's own home. Home ownership in UK is amongst the highest in Europe and there is an antipathy toward renting. Much of this attitude may be explained by an expectation of appreciation in capital value of the abode. There had been substantial private housebuilding developments in the interwar and postwar decades as soon as 'building licence' controls, introduced to deal with materials shortages, had been rescinded. The related public housing programmes have already been mentioned and their demise due to the Ronan Point disaster. This event had repercussions in the private sector. There were still 'slums' to be cleared but people had lost faith in re-development. A way of providing more public housing very quickly was through 'municipalisation'.

Municipalisation involved the purchase by local authorities, on the open market, of existing housing. Often this was older housing in poor condition and located in inner urban areas requiring improvement. Sometimes these would be large houses, 'villas' suitable for conversion to a number of smaller flats; sometimes houses which owners were unwilling or unable to maintain appropriately. Thus councils would be able to increase their housing stock and at the same time take action to improve housing conditions in central areas otherwise in danger of serious decline. Municipalisation was a significant development of the 1970s, when councils were otherwise unable to build. A number of housing associations intervened similarly in these older areas; indeed, the Housing Corporation, the government-sponsored body set up to assist them, classified housing associations as primarily 'new build' or 'rehab'.

Private housebuilding continued apace. When, in the 1980s, a situation was reached whereby there was at last a nett surplus of dwellings over households, considerations became more of housing quality than quantity. With increases in national wealth and disposable income, more people were able to contemplate buying their own home. Home ownership in Britain increased significantly in the 1980s through the Conservative government's introduction of 'right to buy' legislation. This enabled council tenants throughout the country, irrespective of the political persuasion of their local authority landlord, to purchase the house that they had made their 'home' – in essence, a partial 'privatisation' of council housing. For many people this was new ground; instead of being able to call up the previously reviled 'council', former tenants now had the responsibility of organising maintenance for themselves for the first time. More is said about this in Chapters 4 and 6.

After the take-up of 'right to buy' had peaked, the government sought to further reduce the 'tentacles' of local authority control. Legislation was introduced to permit and encourage the transfer of residual council housing to Registered Social Landlords (RSLs) with procedures defined for Large-Scale Voluntary Transfer (LSVT). Housing associations have largely assumed the social housing roles previously the prerogative of local authorities.

The combined effect of 'municipalisation' and 'right to buy' was to give local authorities a much more diverse housing stock, no longer a number of homogenous 'council estates'. Those estates were now peppered by houses not in council ownership but owner occupied and the stock had been augmented by many one-off purchases of very variable construction and condition spread around the district. There were also the estates of the 1950s and 1960s of non-traditional housing to deal with. Altogether a lot of complications for rational maintenance and improvement policies and programmes, and their implementation.

Non-traditional housing

In the quest for a quantitative 'solution' to the 'housing problem', governments around the world have from time to time promoted programmes aimed at speeding up and improving construction. Pawley (1971) and Vale (1995) give good reviews of such developments.

> 'In Britain ... after World War I ... work was directed at overcoming the shortage of skilled craftsmen which dogged the country for a decade. The chief concern was with the production of permanent, low-cost, subsidised dwellings, so natural durability and simplicity were the chief requirements. The means employed were steel and reinforced concrete, both materials

whose potential as building components had scarcely begun to be developed.' (Pawley, 1971, p.51)

The period following World War II brought further non-traditional or system-building developments. In the UK, prefabricated reinforced concrete (PRC) systems included the Airey House, Cornish Units, Laing Easiform, Unity and Wimpey No-Fines. There were also systems that used lightweight aluminium and plywood components made for aircraft production: Arcon, Hawkesley and Tarran bungalows, the 'prefabs' of popular memory, and the British Iron and Steel Federation (BISF) House designed by Sir Frederick Gibberd, later the architect of Harlow New Town and the Roman Catholic Metropolitan Cathedral of Christ the King in Liverpool. Subsequent developments brought the industrialised building systems of the 1960s and 1970s, designed to standardised modular dimensions. The Labour government set up the National Building Agency to vet systems with a view to securing 'variety reduction' and thereby 'economy of scale'. Products of that era included the tower block and 'slab blocks' of flats and maisonettes. These systems were generally of LPS (Large Panel System) design, with room-sized wall, floor and roof panels, and included Bison Wallframe, Jesperson, Taylor Woodrow Anglian and developments of the Easiform and No-Fines systems.

The significance of these developments is twofold. On the one hand, councils were providing themselves with housing stocks of unfamiliar and untested construction, with unknown maintenance needs; on the other hand, tenants bought some of these dwellings under the 'right to buy' legislation. In both instances there are complications for the care of such buildings. In some cases the difficulties of determining appropriate programmes for maintenance, repair and improvement have resulted in demolition of system-built housing long before the 'slums' they were intended to replace, and which have since become very desirable 'bijou residences'. In the case of low-rise PRC houses, the Conservative government was obliged to introduce legislation and financial provision for ensuring their stability when building societies and banks declined to offer mortgages on what they regarded as 'defective' housing, thus imperilling the progress of 'right to buy'.

Non-residential property

In constructional terms, public sector properties other than housing underwent similar transformations. The heritage of Victorian schools and hospitals (variously labelled as infirmaries or sanatoria) were generally conventional brick or stone structures with slate or tile roofs. Postwar developments brought prefabricated and 'temporary' structures to deal with the 'baby boom' and subsequent raising of the school-leaving age (ROSLA). Prefabricated construction systems used included HORSA (Hutted Operation for the Raising of the School-leaving Age), CLASP (Consortium of Local Authorities Special Project),

MACE (Metropolitan Authorities' Consortium for Education), SCOLA (Second Consortium of Local Authorities) and SEAC (South East Authorities' Consortium). There were also many relocatable 'temporary' buildings purchased or hired.

Water leakage through extensive flat roofs and condensation and related mould growth became recurrent problems. These constructions had been procured through 'normal', professional local authority procedures, albeit that for economies of scale their designs had been developed through consortia. The route to resolution of their maintenance problems was not to be the 'normal' local authority, large-scale programme of repairs but through Local Management of Schools (LMS). LMS brought 'the concept of transferring responsibility for a school's day-to-day expenditure to its head teacher and governors... responsibility for repairs and maintenance being shared between the LEA and each individual school' (Davies & Jones, 1993–4). Thus expenditure on the building has to compete directly with recruitment and retention of staff and the purchase of books. It is not yet clear whether the maintenance of schools has been improved by LMS. 'There is a considerable risk that maintenance problems are being stored up for the future by cutting back on basic maintenance expenditure and not using professional advice' (Davies & Jones, 1993–4).

Interest in the maintenance and management of health buildings received a boost in the 1980s when Derek Rayner, the then Chief Executive of Marks and Spencer, was seconded to the National Health Service to administer a dose of market economics. M&S was at that time a very successful clothes and food retailer, a byword for quality, reliability and service. A review of the health 'estate' was undertaken. Perhaps unsurprisingly, an extensive acreage was identified, much of it attached to mental health institutions, formerly known as lunatic asylums. Many of these asylums had been constructed within leafy surroundings, some of which had subsequently been designated as Green Belt and often in areas under pressure to accommodate new housing in the attractive periphery of built-up areas. Thus if the 'inmates' of these institutions could be released to 'care in the community', then substantial estates could be released for development and substantial income generated to top up hard-pressed public coffers. Much commercial interest was shown in such estates in Hertfordshire and Surrey on the outer edge of London and close by its orbital motorway, the M25. Health authorities were encouraged to undertake 'Mereworth' studies, named after a fictitious NHS estate used as a case study for 'rationalisation' of the estate. A former mental institution at Littlemore near Oxford, for instance, has recently been 'rebadged' as 'St George's Park' and converted to desirable, expensive residences.

Private property

With regard to private property other than housing, the private or commercial sector comprises a great multiplicity of building types, including factories and offices, shops and other retail outlets, hotels, restaurants and the leisure and tourism industries. In the UK, prior to the 1950s much of this property was owned and managed by the individual firms that operated within them. Then came a substantial property boom.

'The starting gun for the most intense phase of the property boom was fired on the afternoon of November 2nd, 1954. Mr Nigel Birch, Minister of Works ... announced to the House of Commons that building licences were to be dropped entirely...' (Marriott, 1967). Harold Macmillan, when Minister of Housing and Local Government, introduced the 1953 Town and Country Planning Act which abolished the 100% development charge (the concept introduced by Lewis Silkin in the 1947 Act as 'betterment'): '... the people whom the Government must help are those who do things: the developers, the people who create wealth...' (quoted in Marriott, 1967).

With the property boom came developers like Jack Cotton, Charles Clore and Harry Hyams and firms like Ravenseft and Harold Samuel's Land Securities. Insurance companies such as Legal & General, Norwich Union and Pearl Assurance were looking for investment opportunities and agents like Edward Erdman, Healey & Baker and Hillier Parker were keen to assist in land assembly. Many millions of square feet of new offices and shopping centres were developed in blitzed provincial city centres and the new towns, as well as in London. There were thus developed 'property portfolios' to be managed professionally primarily for profit realisation on the part of landlords and their investors. The tenants, often multiple or chain stores such as Boots the Chemists, British Home Stores and Tesco renting outlets in many of these developments located around the country, also needed their property responsibilities to be managed and effectively maintained so that huge repair obligations would not build up. These were also significant drivers towards the generation of planned maintenance programmes.

The negotiation of appropriate lease terms that determined who was responsible for charges related to management and maintenance of these shops and offices was important to long-term success. A study carried out to investigate planned maintenance practices in the retail sector revealed innovative practice that the authors christened 'just-in-time maintenance' (Smyth & Wood, 1995). More is written on this practice in Chapter 3.

Planned maintenance and the command economy

Against a background of large programmes of housebuilding, new schools and hospitals, with a substantial involvement of 'the state', it is not surprising that maintenance should be approached in a similar, large-scale, depersonalised kind of a way.

In the same way that shortages of labour after wars gave rise to interest and investment in building in ways that used less labour, and also less skilled and unskilled labour, so has there been interest in facilitating effective maintenance with use of minimal resources. By contrast, however, with new construction, the buildings to be maintained are those bequeathed by previous generations and spared the bomber or the bulldozer. In the interwar years, many of the largely similarly constructed houses of the Victorian and Edwardian eras were in individual ownership. Maintenance was ordered by 'landed gentry' or their agents familiar with concepts of good stewardship, and with staff and financial resources. Smaller, tightly planned terraced dwellings were privately rented from landlords who may not have been much interested in their maintenance. Houses built new in the interwar period were not generally much in need of maintenance, other than redecoration, until the postwar period. Schools, hospitals and other public buildings generally had caretakers who would carry out small-scale repairs and report on needs for more major works.

The postwar years brought housing and schools of more variable construction and by the 1970s local authorities were faced with increasing, and increasingly difficult and diverse, maintenance needs. While teachers (prior to LMS) and nurses had little alternative but to put up with the service provided, tenants no longer prepared to put up with poor service were able to move out or to buy. The response of local authorities to this situation was to develop large-scale, comprehensive, multi-million pound rolling programmes of rehabilitation and upgrading.

'Think big'

The bigger, the better seemed to be the maxim of the 1950s and 1960s; Schumacher's *Small is Beautiful* was not published until 1973. Problems associated with programmes have been discussed earlier in the chapter. Size-related problems include the relative 'anonymity' of those creating and those implementing the programme, and their 'distance' from those intended to be the 'beneficiaries' of the programme. Housing officers, planners, architects and their supporting professionals were well versed and experienced in all the conceptualisation and developmental skills needed to design large programmes, new towns and comprehensive redevelopments of older towns; it was an easy step to transfer those skills to the rehabilitation of older housing. After the Ronan Point tragedy, the UK's National Building Agency switched its emphasis from vetting of 'industrialised building' systems to using the same project planning and management techniques for the comprehensive rehabilitation of Britain's older housing. Critical path analysis, 'line of balance' and 'precedence diagrams' were as applicable to rehabilitating thousands of older homes as they were to building thousands of new homes.

However, the Ronan Point disaster extended beyond the demise of the tower block and of industrialised building; there was a demand that housing provision for the future should take more account of what 'the public' actually wanted, not just what 'the professionals' thought they wanted or should have. There developed an expectation of 'public participation' as recommended by the Skeffington Report (1969), together with residents' action groups and 'community architects' such as Rod Hackney, later to become President of the

Royal Institute of British Architects (RIBA), who led a successful campaign to retain and improve the Black Road area of Macclesfield (a small town in Cheshire on the urban periphery of Manchester, UK), where he was living. It has become more complicated to consider who is 'the client': a group of people resident in the area at the time (in which case how to define the boundaries of 'the area'); landlords as well as tenants; elected councillors for the area or the whole council? How should the interests of possible future occupants be taken into account? It may be more common today to speak in terms of 'stakeholders': building owners, investors, landlords, tenants, occupants, users, passers-by, the general public, the parish council, the district council, the county council, central government, pressure groups of various kinds, heritage bodies, Friends of the Earth, international organisations ...

Business and the growth of FM

Arguably, business has long been used to recognising and reconciling the needs, wants and demands of various stakeholders, albeit that the primary object of business is usually to maximise profits for the shareholders. Over the postwar years the route to business success has taken a range of courses, including takeovers and mergers, diversification, consolidation, de-mergers, co-operatives and various stock-option and profit-sharing schemes. More recently, trends have been towards greater focus, with a 'return to core business' accompanied by 'outsourcing' of non-core activities. It is argued that hospitals should focus on surgical operations, not on 'hotel services' such as laundry and the preparation and serving of meals.

The growth of the new discipline of facilities management (FM) was a byproduct of the response to recession of the 1990s. As organisations have sought to return to core business or privatise many operations, facilities management functions have been subcontracted or outsourced. Hand in hand with this has gone a shift in the way in which facilities management, and maintenance in particular, has been defined and evaluated by the client organisation.

These shifts in business approach have been both driven and facilitated by an increasing focus on client needs and customer care.

> 'Understanding needs and delivering promises is not being achieved through current management practices. Nor is it achievable, according to the Nordic School of Marketing (see for example Gronroos, 1990, 1991, 1994; Gummerson, 1994; Storbacka *et al*, 1994), through the traditional marketing mix strategies, using price, product, place and promotion as the key factors, but rather through relationship marketing. The purpose is to increase repeat business and increase profitability.' (Wood & Smyth, 1996).

More is said about marketing in Chapter 4 and about customer care in Chapter 6.

Technology advances

At the same time as management techniques and priorities have been changing, there have been very significant advances in technology. It is an unusual business today, in the Western world at least, that does not make quite extensive use of technology. From electric light and power we have progressed through the telephone and fax to mainframe and desktop computing and information technology and more recently the World-Wide Web and email. As well as speeding up communications and data processing, this has in turn made our buildings and their servicing requirements more complicated.

The kind of office building created in the property boom of the 1950s and 1960s could not anticipate the scale of the communications and information technology revolution. On the whole, office buildings of that era were constructed with minimal floor-to-floor heights and service risers in order to maximise the total lettable floor area within a building envelope whose dimensions were constrained by 'plot ratios' and height limitations imposed by planning authorities. Telephones would be used sparingly; there was a waiting list for telephone lines from the nationalised General Post Office, so often lines and phones would be shared. Most communications with the outside world would be by post and internally to the organisation by memorandum. Consequently there would be extensive banks of four-drawer filing cabinets. Sometimes this would be a constraint on use of the building. For instance, there was a court case relating to the load-bearing capacity of parts of floors of a speculative office building, Tolworth Towers on the Kingston Bypass. Computing, if any, would be done by a mainframe computer of substantial physical dimensions usually housed in a special 'clean room' with a controlled environment. Apart from that, air conditioning was rare, with most offices ventilated naturally through opening windows.

As computing power increased and the size of computers reduced to the microcomputer, so it became possible and then essential to provide desktop computers, one per desk. This, with the consequent need to remove heat from the microprocessors and higher levels of lighting for the computing tasks, generated demand for forced ventilation or air conditioning. Wiring for the electrical supply and distribution, telephones and computers, and ducts for the air conditioning necessitated raised access floors and suspended ceilings to provide service zones for the highly-serviced or 'wired' building.

More recent developments have challenged the air-conditioned approach. 'Wet cooling towers' for cooling and recirculating extracted air have been held responsible for a number of instances of death by Legionnaire's disease. People are also concerned about sick building syndrome (SBS); it is thought that the indoor air quality of some buildings may be a factor. Clements-Croome (2000) includes a number of excellent papers on the interrelationship of environmental conditions in buildings and the productivity and health of occupants of the workplace.

Technology has now moved on again. Combining miniaturised environmental sensors and activators using wire-less radio or infra-red monitoring, with a return to individual control as facilitated by the openable window, has enabled a retreat from the air-conditioned 'solution'. The developing communications infrastructure is also facilitating new ways of working and new conceptualisations of buildings and building types configured around such as the cybercafé, hotdesking, hoteling and tele-cottaging; more on these developments, including automated and intelligent building 'solutions', in Chapter 7 and beyond.

Technology bad, environment and sustainability good

Parallel with developments in technology has been a growing interest in green issues, a concern about the future of the planet, its resources and their depletion. The most commonly cited definition of sustainability is that of the World Commission on Environment and Development (WCED), also known as the Brundtland Report, in 1989: 'the principle that economic growth can and should be managed so that natural resources be used in such a way that the quality of life of future generations is secured'. Works on sustainable architecture or building or construction have been produced by many (e.g. Vale & Vale, 1975, 1991; Halliday, 1994; Anink *et al.*, 1996; Shiers & Howard, 1996).

These concerns are informing new concepts of construction to be measured against a different set of values and criteria. Robert and Brenda Vale developed the concept of the autonomous house, a dwelling that would generate or capture all its own, limited, energy and water requirements and deal with all its own 'wastes' without discharge to the wider environment. A house designed and lived in by Professor Susan Roaf in Oxford, UK, is recorded as attaining a 'zero energy' situation, i.e. it takes no electricity from the National Grid; indeed a small amount of electricity is 'exported' to the Grid from the photovoltaic cells mounted on the south-facing roofslope.

For some, the truly 'intelligent' building would be a 'green' building (Wood, 1999). There are many 'constructions' that can be made of the environmental or sustainable building; there is as yet no consensus. The maintenance requirements of a building will be an important element in determining the consumption of resources over the building's lifetime; its design and specification will be critical to the possibility of adaptation of the building or its component parts to new uses or locations. Alternative constructions, including the maintenance-free building and LEFT and RIGHT concepts (Wood, 2000a,b), are discussed in more detail in Chapter 9.

Summary

This chapter has surveyed a broad sweep of issues related to building conditions and the wider context in which maintenance takes place, or doesn't. Motivating and demotivating factors have been identified. The next chapter discusses what has become the almost standard, professional solution to what are perceived as 'maintenance problems', i.e. planned and so-called preventive maintenance. The pervading rationale is reviewed critically and challenged. Subsequent chapters examine alternative approaches.

References

Anink, D., Boonstra, C. & Mak, J. (1996) *Handbook of Sustainable Building*. James & James, London.

Barbour Index (1998) *The Building Maintenance and Refurbishment Market: Summary*. Barbour Index, Windsor.

Building Design (1998) Refurb market booms. *Building Design* **26 June**, 4.

Burns, R. (1786) *To a Mouse*.

Cahill, P. & Kirkman, J. (1994) Home emergency services. In: *Encouraging Housing Maintenance in the Private Sector* (Leather, P. & Mackintosh, S., eds). Occasional Paper 14. SAUS, Bristol.

Chartered Institute of Public Finance and Accountancy (CIPFA) (1983) *Housing Management and Maintenance Statistics 1982–83 Actuals*. CIPFA, London.

Clements-Croome, D. (ed.) (2000) *Creating the Productive Workplace*. E. & F.N. Spon, London.

Coke, E., (1628) *The Third Part of the Institutes of the Laws of England*, ch.73, p.162.

Confederation of British Industry (1985) *The Fabric of the Nation: a Further Report*. CBI, London.

Damen, T., Quah, L.K. & van Egmond, H.C.M. (1998) Improving the art and science of condition-based maintenance systems. In: *Facilities Management and Maintenance: The Way Ahead into the Millennium* (Quah, L.K., ed.). Proceedings of the International Symposium on Management, Maintenance and Modernisation of Building Facilities, CIB Working Commission W70, 18–20 November, Singapore, p. 114.

Davidson, M. (1995) The English House Condition Survey: past, present and future. *Structural Survey* **13** (4), 28–29.

Davies, H. & Jones, B. (1993–4) Attention all surveyors: our schools need you! *Structural Survey* **12** (5), 31–34.

Department of Education and Science (1985) *Maintenance and Renewal in Educational Buildings – Needs and Priorities*. Design Note 40. Architects and Building Group, London.

Department of the Environment, Transport and the Regions (1998a) *Combating Cowboy Builders: A Consultation Paper*. DETR, London.

Department of the Environment, Transport and the Regions (1998b) *English House Condition Survey, 1996*. DETR, London.

Georgiou, J., Love, P.E.D. & Smith, J. (2000) A review of builder registration in the state of Victoria, Australia. *Structural Survey* **18** (1), 38–46.

Gronroos, C. (1990) *Service Management and Marketing: Managing the Moments of Truth in Service Competition*. Lexington Books, Massachusetts.
Gronroos, C. (1991) The marketing strategy continuum: towards a marketing concept for the 1990s. *Management Decision* **29** (1), 7–13.
Gronroos, C. (1994) From marketing mix to relationship marketing: towards a paradigm shift in marketing. *Management Decision* **32** (2), 4–20.
Gummerson, E. (1994) Making relationship marketing operational. *International Journal of Service Industry Management* **5** (5), 5–20.
Halliday, S. (1994) *Environmental Code of Practice for Buildings and their Services*. BSRIA, Bracknell.
Inside Housing (2000) DETR and Treasury scrap over backlog. *Inside Housing* **17**(15), 1.
Macmillan, H. (1957) Speech at Bedford, 20 July 1957. *The Times* 22 July.
Macmillan, H. (1985) Speech on privatisation to the Tory Reform Group, 8 November 1985. *The Times* 9 November.
Marriott, O. (1967) *The Property Boom*. Hamish Hamilton/Pan Books, London.
Milne, R.D. (1985) *Building Estate Maintenance Administration*. E & F.N. Spon, London.
Ministry of Housing and Local Government (1966) *The Deeplish Study*. MOHLG, London.
NHS Estates (1995) *Estatecode: Information for the Facilities Management Function*. HMSO, London.
Pawley, M. (1971) *Architecture Versus Housing*. Studio Vista, London, p.24.
Pitt, T.J. (1997) Data requirements for the prioritisation of predictive building maintenance. *Facilities* **15** (3/4), 97–104.
Rolfe, S. & Leather, P. (1995) *Quality Repairs: Improving the Efficiency of the Housing Repair and Maintenance Industry*. Policy Press in association with the Joseph Rowntree Foundation, Bristol.
Society of Chief Architects of Local Authorities (1993) *Maintenance Expenditure '92*. SCALA, Nottingham.
Schumacher, E.F. (1973) *Small is Beautiful: A Study of Economics as if People Mattered*. Blond & Briggs, Abacus/Sphere Books, Falmouth.
Seeley, I.H. (1976) *Building Maintenance*. Macmillan, London.
Shen, Q. (1997) A comparative study of priority setting methods for planned maintenance of public buildings. *Facilities* **15** (12/13), 331–339.
Shiers, D. & Howard, N. (1996) *The Green Guide to Specification*. Post Office, London.
Skeffington, A. (1969) *People and Planning: Report of the Committee on Public Participation in Planning*. HMSO, London.
Smyth, H.J. & Wood, B.R. (1995) *Just in Time Maintenance*. Proceedings of COBRA '95: RICS Construction and Building Research Conference, Vol. 2, pp. 115–122. RICS, London.
Spedding, A. (ed.) (1994) *CIOB Handbook of Facilities Management*. Longman, Harlow.
Storbacka, K., Strandvik, T. & Gronroos, C. (1994) Managing customer relationships for profit: the dynamics of relationship quality. *International Journal of Service Industry Management* **5** (5), 21–38.
Thatcher, M. (1987) in *Woman's Own*, 31 October.
Then, D.S.S. (1995) Computer-aided building condition survey. *Facilities* **13** (7), 23–27.
Vale, B. (1995) *Prefabs: A history of the UK Temporary Housing Programme*. E. & F.N. Spon, London.
Vale, B. & Vale, R. (1975) *The Autonomous House: Design and Planning for Self-Sufficiency*. Thames and Hudson, London.

Vale, B. & Vale, R. (1991) *Green Architecture: Design for a Sustainable Future*. Thames and Hudson, London.

Wood, B.R. (1986a) Surveying the estates. *Housing* (Journal of the Chartered Institute of Housing) **22**(2), 24–25.

Wood, B.R. (1986b) The condition of local authority housing. *Housing and Planning Review* **41**(3), 7–8.

Wood, B.R. (1998) *Maintenance Service Development*. Proceedings of COBRA '98: RICS Construction and Building Research Conference, Oxford Brookes University, 2–3 September. RICS, London.

Wood, B.R. (1999) *Sustainable Building Maintenance*. Proceedings of Catalyst '99, University of Western Sydney, 5–7 July, pp. 129–140.

Wood, B.R. (2000a) *Sustainability and the RIGHT/ LEFT Building*. Proceedings of Conseil Internationale du Batiment Joint Symposium of Working Commissions W55 and W65, University of Reading, 13–15 September.

Wood, B.R. (2000b) *Sustainable Building Care*. Proceedings of Conseil Internationale du Batiment Working Commission W70 Symposium, Queensland University of Technology, Brisbane, 15–17 November.

Wood, B.R. & Smyth, H.J. (1996) *Construction Market Entry and Development: The Case of Just in Time Maintenance*. Proceedings of 1st National Construction Marketing Conference, Centre for Construction Marketing, Oxford Brookes University, July.

World Bank (1996) World Bank supports enterprise restructuring through housing sector reforms in Russia. Press Release No. 96/36 ECA, 7 May, 1996. (www.worldbank/projects/portfolio/habitation96.htm)

World Commission on Environment and Development (WCED) (1989) *Our Common Future*. Oxford University Press, Oxford.

2 Planned Preventive Maintenance Prevails

Planned maintenance programmes have been *de rigeur* for many years now. The rationale is that it should be possible to anticipate maintenance requirements and, by planning timely interventions, to achieve economies of scale. It is also axiomatic that by considering the performance of materials and components over time it is possible to identify the point at which a maintenance operation, be it of repair or replacement, could be executed prior to a failure of the component. In this way the maintenance operation could also be said to be preventive. This chapter casts a critical eye over the dominant ideology of planned preventive maintenance (PPM) regimes.

Some definitions

Definitions are not static; they evolve and change over time in the light of experience and re-evaluation.

Maintenance is 'a combination of any actions carried out to retain an item in, or restore it to an acceptable condition' (BS 3811: 1984).

BS 8210: 1986 defines 'building maintenance' as 'work, other than daily and routine cleaning, necessary to maintain the performance of the building fabric and its services'.

BS 3811: 1993 refers to BS 4778, Part 3, Section 3.2: 1991 for its definition of maintenance as 'the combination of all technical and administrative actions, including supervision actions, intended to retain an item in, or restore it to, a state in which it can perform a required function'.

Planned maintenance was defined by BS 3811: 1984 as 'maintenance organised and carried out with forethought, control and use of records to a predetermined plan'. BS 8210: 1986 conditions this by adding (after 'plan') '. . . based on the results of previous condition surveys'.

Preventive maintenance is 'maintenance carried out at predetermined intervals or to other prescribed criteria and intended to reduce the likelihood of an item not meeting an acceptable condition' (BS 3811; 1984). Later BS 4778, Part 3, Section 3.2: 1991 amends this to 'the maintenance carried out at predetermined intervals or according to prescribed criteria and intended to reduce the probability of failure or the degradation of the functioning of an item'.

The antithesis of planned maintenance is reactive or response maintenance and of preventive maintenance is breakdown or emergency maintenance.

Provenance

The background to the development of planned preventive maintenance (PPM) is of large-scale programmes of repair, maintenance and improvement of large stocks of buildings, particularly in the public sector. They have become 'received wisdom' and as such are relatively unchallenged as the 'obvious' or 'professional' approach to maintenance. The virtues were discussed by many in the 1970s and '1980s and with official support, for instance Bushell (1979), the Scottish Local Authorities Special Housing Group (1979) and the Housing Services Advisory Group (1980). The rationale was related to thinking and operating on a grand scale and was a logical extension of the kind of thinking that had dealt with slum clearance, the rebuilding of large areas flattened by bombing and the Blitz, and the planning and construction of complete new towns. It fitted in with organisations predicated on 'public service' and departmental structures of central and local government supported by a complement of administrative, professional and technical staff well used to working within defined set of procedures - a 'reductionist' kind of approach. Everyone knew their job, with defined limits of responsibility and authority; there was a 'right' way to do things, a documented procedure, leaving little to the imagination and little to go wrong.

The advantage of prevention is almost too obvious to require assertion: 'a stitch in time saves nine'. The avoidance of failure is a desirable attribute, but at what cost is this to be achieved? Failure of a component may result in equipment breakdown with consequential inability to provide service. The consequences of failure may be expensive and in some circumstances catastrophic. Failure could almost certainly be completely avoided if all components were replaced almost immediately after installation, although even then there is the chance of installing a defective component or of creating some other fault by inadvertently interfering with otherwise satisfactorily functioning components. It is important therefore to be able to identify the optimum time or condition at which a 'preventive' intervention should be made. Unfortunately, reliable and comprehensive data on component performance and criteria on which to form a judgement of the optimal time are often hard to find.

Automaticity

An advantage of the PPM approach is that decision making becomes straightforward, perhaps giving rise to 'automatic' responses, i.e. 'if ... (this situation), then ... (that response)'. (*Note:* this *automatic* response is not to be confused with an *automated* response, that is one implemented using automation, such as a robotic device. Issues related to automation and automaticity are discussed further in Chapter 7.) Thus, for instance, it may be decided that whenever windows come due for redecoration they will instead be replaced in 'maintenance-free' PVCu to a standard design supplied at a 'good price' negotiated through a bulk purchase arrangement. Where any work is required to a rendered chimney stack, it should be held in abeyance (unless a recognised emergency) pending rebuilding to a standard detail in approved brickwork.

A danger of this approach is that there may be a tendency to make decisions 'at the desk' or on the phone, in front of the screen that displays the 'answer' to the symptoms entered into the computer, rather than to visit the property to assess the situation 'in the field'. A site visit, though time consuming, allows a fuller consideration of the context; it may be more likely that a full, professional view is obtained thereby and that thus the correct, effective repair is ordered up. Both surveyor and building user may also benefit from incidental discussion at the time of the visit. This may allow either for 'education' of the occupant about how better to use the building or to justify why a particular repair action is proposed; it may also allow for meaningful discussion about best timing, consequential disruption, etc. Against this, though, must be placed some reservation about surveyors' technical skills as reported by Hollis & Bright (1999).

It may be that surveyors' time would be best invested in devising effective decision support systems. Software packages exist that are claimed to provide such 'maintenance solutions', with icons, drop-down menus, prepared scripts with questions to ask, and decision trees that through a structured 'interview' arrive at the 'right answer'. Perhaps such systems could be customised to particular estates.

Fail to plan; plan to fail

This expression sums up so eloquently and succinctly much of the logic behind the development of PPM, and the 'comfort zone' attained by its proponents and practitioners; I would cite the provenance of this quotation if I could attribute it. It encapsulates the superior logic and professionalism that effective planning represents. It suggests that every eventuality is identified, all possible ramifications fully evaluated and every alternative response, including inaction, analysed and assessed, including possible 'side-effects', and associated risks fully considered in the equation. It suggests the possibility of 'calculating' the right answer; the term 'calculated risk' is associated.

The 'planned' approach is one with which many feel comfortable. Maintenance has not historically been an area of activity known for being populated by risk-takers; safety first, reliability and avoidance of breakdown are watchwords. Failure is something to be avoided, possibly 'at all cost', and the price of failure may well be loss of confidence in an organisation or an individual and possibly loss of job or even loss of life.

Computers and LAMSAC

Local authorities and public bodies in Britain in general are used to the idea of sharing expertise and experience, of promulgating good practice and learning from the mistakes, misgivings and criticisms of their peers, fellow professionals in similar circumstances. These authorities are also accountable to what used to be known as ratepayers (later Community Charge payers and now Council Tax payers) and to central government and ultimately to the general public, taxpayers and the electorate. Britain has a long heritage of accountability for the expenditure of public funds.

With the advent and advancement of computers within local government there was a particular need for professionals to be able to decide on ways of making best use of their capabilities. Anything which smacked of big numbers and/or big expenditure was worthy of serious consideration for 'computerisation' and the need to make the right decision about the best way forward was critical. The price of failure, of buying the wrong system, of losing all the rent records, of spending a lot of money on something that broke down or failed to deliver was very high. IBM, ICL, Honeywell and McDonnell Douglas were the main providers of hardware but what about software, and maintenance?

In the same way that authorities had banded together in consortia to commission design and construction of system-built flats or schools, they recognised the value of commissioning advice on computerised systems – for rent and rates accounting, for processing invoices and payments, for repair requests and planned maintenance. The Local Authorities Management Services and Computer Committee (LAMSAC), for instance, produced in 1980 authoritative guidance on the benefits of computer systems particularly for housing maintenance and repairs.

Computers have been used to assist the maintenance management process since the early 1970s and by the mid-1980s many maintenance organisations were using software developed for large mainframe computer systems (Pettit, 1983).

> 'The software was normally designed around a central computerised database into which maintenance and repair information was recorded. The information was then manipulated to produce works schedules and job orders. Report generators allowed work-in-progress to be monitored and statistical management information to be produced.' (Jones & Collis, 1996)

Audit Commission and the 3 Es

The Audit Commission exists to assess the performance of public bodies, particularly local authorities and the National Health Service, and to assist them to obtain better value for money through the identification and dissemination of good practice. Typically, it assembles data and undertakes studies of current practice, which it then publishes for the benefit of public authorities more widely. Through the 1980s the Commission pursued and promulgated policies aimed at achieving what it referred to as the 3 Es.

- Economy
- Efficiency
- Effectiveness

It is important, while pursuing these goals, to recognise the differences of emphasis. Drucker (1967):

> 'pointed out that to concentrate on efficiency rather than effectiveness could be limiting and therefore dangerous. It can result in:
>
> - doing things right rather than doing the right things
> - solving problems rather than producing creative alternatives
> - safeguarding resources rather than optimising resource utilisation
> - lowering costs rather than increasing profit
> - minimising risk rather than maximising opportunities.'
>
> (Armstrong, 1994)

In respect of economy, the danger is the pursuit of cheapness or false economy. In the case of tendering, in theory at least, all tenderers are submitting their prices against a consistent and complete specification of quantities and quality, the differences in tender prices being a measure of relative efficiencies of the firms. Unfortunately, it may represent no more than underestimation of complexity and miscalculation and/or an expectation of 'cutting corners' in the work. False economy results where a cheap repair job is done when a more expensive repair or replacement may achieve much better longevity; however, if insufficient funds are available there may be little real choice.

Studies by the Audit Commission of local authority maintenance practices (1986a,b) together with analyses of costs from the Chartered Institute of Public Finance and Accountancy (CIPFA, 1983) informed the article (Wood, 1986) referred to in Chapter 1. That highlighted the disparity in spending between Kingston-upon-Thames and Kingston-upon-Hull councils on maintaining their housing stocks. Another study showed that Wolverhampton Metropolitan Borough Council, a large local authority in England's West Midlands conurbation, had an external redecoration cycle of 24 years, i.e. that if it only repainted annually the number achieved in the year under review, it would take 24 years to complete the cycle.

Audit Commission studies focus on reducing costs, obtaining better value and managing more effectively and have included: Saving Energy in Local Government Buildings (1985), Managing the Crisis in Council Housing (1986) and Property Management in Local Authorities (1987).

Condition surveys and elemental repairs and replacements

The need for condition surveys was identified by BS 8210: 1986 as a prerequisite for planned maintenance. Discussing the rationale of investing in building condition surveys, Then (1995) states that 'A condition survey is a form of building inspection. However, unlike routine inspections and planned inspections, which are geared towards the issue of work instructions, condition surveys usually take on a longer-term view'. It enables a strategic overview of the building stock and facilitates discussion and decisions on policy and priorities.

With this in mind, condition surveys are therefore normally of a rather 'generalist' flavour. In relation, for instance, to local authority or housing association dwellings, a sample of properties will be inspected, sufficient to derive a budget and programme of work for the whole. It is important to select an appropriate sample, generally one of each discernibly different plan type and construction per estate. There may be different elevational treatments, for instance some of cavity brickwork, some half-rendered and some fully rendered; some may have hipped roofs and some gabled; some slated and some tiled. The important thing is to be able to deduce a reasonable estimate of the work required to the whole by inspection of a sample. Normally condition is the determinant of appropriate action and its priority for external elements, whereas internally the prime consideration is more of deficiencies or poor plan arrangement rather than defects. For instance, the number and disposition of electrical outlets, especially in the kitchen, and the arrangement of the bathroom and sanitary accommodation are more common problems than defective plasterwork or dangerous wiring, although of course those also exist and require action.

Often a 10% sample will suffice; perhaps even smaller on the usually more repetitious interwar estates and generally larger on the post-1970 developments where dwellings tended to be assembled into more varied arrangements. A tower block may have few plan variants, but it will be important to survey flats on different floors and on different elevations to reflect different exposures to sun, wind and rain, the principal agents of deterioration and decay. Lomas (1997) gives a good resumé of possible survey problems relative to size of survey team. The survey objective must be kept in mind; properties are selected and surveyed only as much as required to determine approximate totals and priorities.

Every condition record and action assessment made in the field is capable of being overridden in the final analysis when elements are considered in relation to one another and in total and in the quest for assembly of large 'packages' of related works. Economies of scale and attractive contracts would be expected concomitants. All assessments of repair or replacement and their anticipated timescales are provisional pending this overview.

> 'The condition is rarely the only consideration; other considerations such as the logic of undertaking work at certain times of the year or combining tasks

that require common temporary works such as scaffolding will influence priority. Also political considerations might be an influence.' (Pitt, 1997)

Virtues of the 'elemental' approach are that surveyors are able to create a fuller picture of the property and the aspects more in need of attention and that programmes can give priority to those elements most in need of prompt attention overall. It is also helpful in that 'Budgets for ... building maintenance in many countries cannot meet the ever-increasing maintenance needs...' (Shen, 1997) and because resources are never sufficient to 'solve' problems of maintenance backlogs (Then, 1995).

Programmes and professional service

The production and implementation of PPM programmes are obviously 'professional' activities. The condition surveys need to be planned and undertaken by people with appropriate expertise, normally surveyors of some kind, maybe chartered, perhaps architects or technician staff. Analysis and assessment of the data collected are also professional tasks, as are the preparation and balancing of budgets and programmes. The letting and administration of contracts, typically for several hundreds of thousands or millions of pounds, euros or dollars, also warrant professional attention.

By contrast, the repair of a single broken or malfunctioning component seems a lowly, almost inconsequential job; all that is required is for someone to attend to 'fix it'. Therein may lie a number of problems. A 'running repair' to keep a piece of equipment operating satisfactorily may suffice for the time being; it may then be merely a matter of time before the next failure. A 'like for like' replacement component may have the advantage of simplicity, especially if components are modular or standard, and 'in stock'. Both these approaches require recognition of the defective item to be repaired or replaced; both, however, also run the risk of dealing only with the *symptoms* and failing to investigate or identify the underlying *cause*. In the event that the work is carried out by a contractor unfamiliar with the client or the premises, called out with no brief beyond 'fix it', it is likely that the defect will not be tracked back. Little learning will be achieved from which to benefit on the next occasion.

When a surveyor is involved there is a professional obligation to 'follow the trail' of a defect back to the root cause or at least to recommend where further investigation may be warranted. Oxley (1998) highlights a County Court decision (*The Alexander Collections Ltd* v. *Martin and Lacey*) where the judge criticised the defendant for not following the 'trail of suspicion'. The surveyor's report did not clearly reflect the nature and extent of the dangerous situation. From another case, *Oswald* v. *Countrywide Surveyors Ltd*, Oxley (1998) concludes that a 'recommendation for further investigation would suffice in most cases due to the limitations usually encountered during inspections for survey and/ or valuation purposes'.

As councils and housing associations only commission condition surveys around every five years or so, and private buyers very much less frequently on average, finding a good and reputable surveyor is very important. 'Most clients requiring a structural or other condition survey of a building are infrequent consumers of such a service' (Hoxley, 1995). Hoxley identifies from a number of sources (Wittreich, 1966; Zeithaml, 1981; Wilson, 1984; Connor & Davidson, 1985; Wheiler, 1987) the importance of referrals; he quotes Wilson thus: 'In the ... professions estimates of all new business from referral range from 80–100 per cent'. Hoxley also quotes Wheiler quoting Donnelly & George (1981).

> 'The professional is supposed to be an authority on his subject, as a body of knowledge, and an expert on its application to the solution of particular problems presented to him by the client. The client is thus placed in a subordinate position in the relationship with the professional.'

Thus, through the execution of condition surveys and the preparation of maintenance and repair programmes, the surveyor obtains benefit from continuity of work and the client benefits from avoiding the risks and uncertainties associated with working with unknown and untried personnel; a desirable 'win–win' situation.

Communications and information technology

Mention has already been made of the introduction of computing into maintenance management in the 1970s and 1980s, particularly in local authorities (Pettit, 1983; Then, 1995; Jones & Collis, 1996; Pitt, 1997; Shen, 1997). In the 'early days' computing was virtually synonymous with 'data processing', largely a matter of assembling and manipulating large quantities of data and making little use of the computer's ability to 'think', to consider the 'what if...?' possibilities. 'All of the early computerised maintenance management systems computerised existing manual procedures, thus building in unnecessary constraints' (Pitt, 1997). With the move from mainframe to desktop, 'many software houses developed property/maintenance management packages ... more comprehensive in nature, using relational database development tools...' (Jones & Collis, 1996). Database software has become increasingly cheap and powerful, enabling condition data to be readily analysed and available at the maintenance manager's desktop (Then, 1995), though this is not without problems. Shen (1997) refers to the problem whereby 'since maintenance budgets ... are significantly less than the real maintenance needs, the cut-off line usually falls somewhere in the middle of a priority category'. A response to this is to enable differentiation by creation of more categories or to identify conditional relationships.

The ability of computer software to accept and manipulate multi-factor criteria has been described and developed by a number of authors: Tsang, 1995; Jardine *et al.*, 1997; Pitt, 1997; Shen, 1997; Triantaphyllou *et al.*, 1997. These often develop earlier work in the fields of statistical analysis, especially with reference to reliability and risk, and aimed at assisting related decision making. References are made in particular to:

- analytic hierarchy process (AHP) (Saaty, 1980)
- proportional hazards modelling (PHM) (Cox & Oakes, 1984)
- multi-attribute utility theory (Finch, 1988)
- reliability-centred maintenance (RCM) (Moubray, 1990)
- intelligent maintenance optimisation (IMDS) (Kobbacy *et al.*, 1995)
- decision theory (Almeida & Bohiris, 1995)
- failure modes and effects analysis (FMEA) (Layzell & Ledbetter, 1997).

Most of the above relate to methods of prioritisation using calculations in which identified criteria are 'weighted' and 'factored' to give an 'answer' that reflects the multiplicity of demands on the maintenance service. Arguably, this would be of less importance if there were enough funds to meet the identified needs.

Shen refers to prioritisation methods used in British local authorities as reported by Spedding *et al.* (1994) and lists six major criteria:

- building status
- physical condition
- importance of usage
- effects on users
- effects on fabrics
- effects on service provision

Pitt, developing this approach, illustrates that 'using this approach with five physical condition categories, three fabric-effect categories and three user-effect categories, there are 45 combined categories'.

The foregoing demonstrates the ability of computers, given appropriate criteria with which to address the data, to produce multiple versions of PPM programmes. This is predicated on having complete and consistent data, related to anticipated repair and replacement actions on all relevant elements on each building in the property 'portfolio'. It presupposes that maintenance activity can be anticipated and preplanned, which may be largely but not entirely the case. It will be an unusual building in which all components perform at or above the desired level until repair or replacement just before the anticipated failure. This will be discussed further in later chapters.

Components or services failing outside a PPM programme will require action of some kind. It may be that repair or replacement can be deferred until the time determined in the PPM programme; perhaps the breakdown may lead to advancement of a related part of the PPM programme. Perhaps a temporary repair may be appropriate, pending replacement in accordance with the programme. There are many questions and the 'answer' will be very much context related. Here too, the computer has much to offer; indeed, many of the earliest maintenance management software programs were developed around the ordering of 'day-to-day' or 'response' maintenance.

Response or emergency maintenance

This was referred to earlier as the antithesis of PPM; it presupposes that planned and preventive maintenance are ideals. Even in the best thought-out programmes there will be breakdowns and response to these can also be planned, albeit on a more 'one-off' basis. Roof leaks, window breakages and flooding are all examples of situations demanding immediate response. In some cases, further investigation may be required to identify the precise remedial work required, and some may need an element of diagnosis to determine the underlying cause. Often, therefore, the work carried out does not accord with that which may have been anticipated at the outset, when the damage or defect was detected. The 'ordering' of maintenance in response to a notification of a defect or a repair request is likely to require several steps to be taken, which may include the following.

(1) Building user identifies damage or defect.
(2) User reports through line management to 'responsible officer'.
(3) Phone call, fax or email to 'maintenance department' or equivalent.
(4) Maintenance clerk identifies likely 'trade' required, e.g. carpenter/joiner.
(5) Clerk identifies forward availability of staff or contractor.
(6) Clerk places 'repair order'.
(7) Maintenance operative attends; identifies that work required is not as stated on order; requests authority to execute work needed or for work to be reprogrammed; perhaps requires roofer, glazier, etc.
(8) Repair eventually executed.
(9) Clerk of works attends for random quality check; identifies further problems...
(10) Back to 'square one'?

The above is not intended to report negatively on response maintenance, only to illustrate some of the complexities and pitfalls of the 'one-off' repair request. These complications have tended to support moves towards maximising the proportion of total maintenance work to be incorporated within PPM programmes. The Audit Commission recommended (1986b, p.28) that 'this percentage needs to rise to about 65–70% of the maintenance spending'. This expectation continues today, as witnessed by 'best value' reports that can be viewed on the Web (www.bestvalueinspections.gov.uk) and in a summary report (Audit Commission, 2001, p.17). This, however, fails to take advantage of the responsive mode and overlooks some of the pitfalls of PPM.

Modern technology has facilitated remote monitoring of components and services and has enabled automatic 'calling up' of maintenance, perhaps to a call centre or help desk. Software permits the immediate ordering and reordering of priorities which, together with paging and mobile or roaming telephony, allows an available operative to be identified and 'despatched' to the job. The carrying of a limited inventory of standard or modular components may enable a breakdown to be avoided rather than just rectified. A 'just-in-time' approach is described in Chapter 3.

Objections to PPM

'You can never plan the future by the past.' (Burke, 1791)

The previous parts of this chapter have focused on the features of PPM and factors conducive to its development and successful implementation. The following section focuses on facets less conducive to PPM and situations in which alternatives to PPM should be more seriously considered.

Considerations militating against PPM include:

- unutilised service life
- user wants and needs
- overspecification and overwork
- unresponsiveness
- resource intensity
- unsustainability
- use of large contractors
- use of non-local labour
- adversarial nature of contracts
- automaticity, detracting from the individual.

Unutilised service life

A consistent objection to PPM has been the replacement of components where 'they haven't failed yet!' The situation arises where, for instance, it has been decided (and sometimes it will be unclear who decided, and why) to replace a whole floor of fluorescent lighting tubes, despite the fact that just one or two on that floor have failed. It may be that, on another floor of the building, a number of tubes have failed and it makes sense from an economic and operational point of view to replace all these in one go. Furthermore, it is possible at this time to upgrade the fittings to be more energy efficient and thereby to save energy and expenditure in the long run.

However justifiable the replacement may be, it is still the case that unutilised service life is being wasted. Until recently such tubes would be consigned to the refuse tip. However, it has been recognised that the fluorescent gases within the tubes contribute to the greenhouse effect and global warming and that they should be disposed of with care. Perhaps a market in second-hand luminaires could be developed, in the same way as recycled CFCs and halon from existing, elderly refrigeration and oxygen-starvation fire-extinguishing systems that still have useful life elsewhere.

In similar vein, window replacement programmes have tended to involve the replacement of all windows in each dwelling on an entire estate –

thousands of windows over a period of years. This is not to say, however, that piecemeal replacement of odd windows as and when required is necessarily ideal either. Picture in your mind the front elevation of a 1920s-built semi-detached house with a steel-framed 'Crittall' window to one of the bedrooms in the original small square-paned pattern, another with the 'sun-trap' rectangularly subdivided lights and the ground floor window in modern plain-glazed lights – an architectural nightmare! At the other end of the repair/replace spectrum, is there not a case for replacing all the windows, and doors too, with double- or even triple-glazed 'maintenance-free' PVCu units? Suffice it to say here that such units are unlikely to be maintenance free. A study by Berry (2000) of such wholesale window replacements in a number of local authority housing stocks has shown a continuous need for attention to the associated ironmongery; easing and adjustment of hinges and attention to locking mechanisms. Is there a use for recycled steel windows?

User wants and needs: putting the customer last?

The all-encompassing logic of the PPM programme tends to give overriding weight to the overall needs of the stock as perceived by the maintenance manager. There are so many considerations and it is his or her job to balance those and to arrive at the answer. It is very difficult in such a process to give any weight of consideration to an individual's desires. Whilst accepting that I may tend to express my own wants or desires as 'needs' and 'essential', it is generally the case that I will only get what I want in a PPM programme if that corresponds with the overview. 'The greatest good for the greatest number' would be a typical approach. The 'professional' is also much more likely to emphasise the technical and operational considerations that may not be important to the building owner or users.

Computing power, however, and the ability to work efficiently within smaller 'batches' than assumed by the stock-wide programme allows the approach to be altered. If desired, it is possible to provide maintenance on a much more individually focused basis. Why should a building tenant not be permitted choice? Perhaps a tenant may prefer to wait for new windows or new bathroom fittings until her daughter's school examinations are completed. Another tenant may prefer a different design of kitchen or fittings of a higher specification.

Would it feel different if tenants were considered as 'customers' or 'clients' and asked what they want? Could there be a charging system that created different bills or differential rents for people selecting different services? Should tenants have less choice, or no choice, because they are tenants; should they 'put up with what they are given'? It is commonly believed that tenants know little and care less about what makes for a good environment and how to exercise choice wisely. In fact, evidence suggests that building occupants are well able to make effective choices and are not profligate or unreasonable in their expectations. Does 'mother know best'? This will be examined further in Chapters 5–7.

Overspecification and overwork: more work?

The PPM programme presupposes 'thinking big'; big is beautiful. This means that where, for instance, all windows but one in a house are deemed by a surveyor as requiring early replacement, it makes sense to replace all the windows at that time. What if the one window, the 'exception', had been replaced already and quite recently? Perhaps the decision on whether to replace that one again now will be taken in anticipation of how many 'exceptional' windows there may be on a particular estate; how much money, if any, will be saved by omitting the window from the programme. It is quite likely that the overview, at programme level, will dictate that the window, which is at present 'satisfactory', should be replaced, for the benefit of future maintainability, efficiency and 'cost-effectiveness'.

There is, however, more work involved in total in taking this approach. The properties as a whole have been surveyed, albeit that sample survey techniques will have reduced the survey load that would be implied by a 100% survey. The determination of the survey sample and the 'plan of attack' for the survey fieldwork will also take time and effort. There is further work involved in data entry, analysis and assessment; the analysis software is not free, nor are the data entry and analysis free of error or breakdown. There is also the determination and application of criteria for assessment of condition and of appropriate repair/replacement/improvement actions. This also supposes that surveying 'professionals' act professionally and apply consistent standards.

Hollis & Bright (1999) show that there is great variability between individual surveyors and that surveyors miss many defects that should be identified. The study followed up on an identified concern about standards being applied when surveyors carry out surveys. Ten surveyors were commissioned to complete a survey of a particular house, using the standard RICS Homebuyer Survey and Valuation pro forma, which has been in use for more than 15 years.

> 'Of the six defect items ... one surveyor identified all six items within the report, one surveyor identified five items and three surveyors identified four items. The remaining five surveyors failed to identify more than two of the items listed and one surveyor did not identify any of the defects.' (Hollis & Bright, 1999)

This surveyor variability, and inadequacy, calls into question the ability of qualified surveyors to consistently identify, interpret and report on defects and what should be done about them. Perhaps there is something to be said for running components to 'failure' and then replacing them, thus cutting out the 'professional' and related decision making which seems otherwise so unreliable. Depending on the need for 'running repairs' or 'servicing' or redecoration, this approach may also have the least number of maintenance interventions and least lifetime cost. This would be worthy of further investigation.

Unresponsiveness

Another criticism of the PPM approach is its relative inflexibility and unresponsiveness to changing circumstances. The habit of thinking 'big' and 'long term' tends to create a 'supertanker' which, because of its size and momentum, is slow to stop or change direction. The context within which the programme is being developed and/or implemented may change quickly and substantially. Changes may be political, economic, social or technological. There may be changes in political control at central or local government level; changes in management personnel, policies and control. National and global economies boom and bust; the fortunes of firms fluctuate, projections are changed, up and down, and budgets increased or decreased. Social and societal changes result in changing demographics and necessitate more accommodation, or less, or differently fitted or configured. Technology seems to advance continually: 'Moore's Law ... named after Intel founder Dr Gordon Moore, states that semiconductor transistor density, and hence performance, doubles roughly every 18 months' (Azar, 2000).

Once a programme is established it is difficult to change. The investment of time and professional commitment to the programme has been recognised. Once a budget has been set it is difficult to cut back. Components will already have been ordered and there will be charges to take them back into store; if they are non-standard, of special size or specification, they may not be returnable or resaleable to another client. It is inconvenient to redeploy or lay off skilled operatives and they will be expensive and hard to find when the programme picks up again when finances improve. Similarly, it is difficult and expensive to accelerate programmes. Sometimes finance may become available from other programmes which have for some reason stalled or failed to progress as planned and budgeted. With regard to budgets in the public sector, there is no worse crime than to not spend or to underspend; this shows an inability to spend and the consequence is to have future budgets cut accordingly. Conversely, the reward for overspending is to have budgets increased as they were clearly inadequate previously!

What tends to happen is that a PPM programme is based on a comprehensive condition survey executed on the basis of perhaps explicit, but more likely implicit, standards and repair/replacement/improvement criteria; and that programme will largely hold good until such time as the next condition survey is commissioned. Thus a ten-year rolling programme may be revisited and reprioritised after five years; it is more likely that one five-year programme will be followed on completion by another, newly constructed from new data and applying new criteria. Thus the programme responds to changed thinking at five-year intervals.

Resource intensity of PPM: simple but heavy on resources

A significant virtue of PPM programmes is their relative simplicity of preparation and operation, though that is bought at a price of high resource use. Expensive professional time is required to execute the requisite condition surveys and to turn the millions of data items into a programme that is simple to understand and implement. In many circumstances the human and/or financial resources to do this may be in short supply or unavailable or only available in the short term at the expense of immediate maintenance needs that will then be deferred.

The execution of the programme may also require either an increased use of resources or their redirection from other work. Typically, the execution of a condition survey identifies much more maintenance, repair and improvement work than had previously been thought necessary; thus new and larger budgets need to be created and these need to be funded from somewhere. Either income, through rents and charges, needs to be increased or other expenditures need to be curtailed or extended over a longer period than previously envisaged. The tendency of PPM programmes to 'override' repair decisions that may be adequate and appropriate for individual situations has been referred to; so too has the need for substantial professional input. To this should be added that 'experience has shown that, frequently, the demand for maintenance, as identified by the condition survey, exceeds the available budget' (Pitt, 1997). Pitt imports a figure from Spedding *et al.* (1994) that shows maintenance needs for a sample of UK local education authorities to be between two and ten times the approved budgets.

There is also the matter of greater use of physical resources already referred to as unutilised service life: windows or fluorescent tubes replaced when still functioning satisfactorily, because it 'makes sense' in programme terms to replace the lot now. It would make an interesting and useful study to identify and quantify this squandering of scarce and in some cases non-renewable resources.

Commonly the mere collation of data from a condition survey results in a summation not previously done or not done so comprehensively. Is it that previous budgets are more likely to have been based on less 'objective' criteria? It is common for a budget to be set at the same level as the previous year with a small percentage change – either plus $x\%$ to allow for inflation or minus $n\%$ representing a cut across all budgets as 'efficiency gains' or some similar euphemism for cuts. There is legitimacy in these approaches; they represent a demonstrated ability to perform at that level and/or a capability to resource at that level. The 'objective' assessment of 'maintenance need' is in fact another expression of a 'desire', albeit a professionally determined one.

The production of a PPM programme and supervision of its implementation are 'high-level' activities; they demand strategic thought, weighing and resolving conflicting demands. This is attractive; it commands the respect due to a professional and has status attached to it. It takes the surveyor 'out of the gutter' and onto the high ground of strategy development. These may be factors in the promotion of PPM. Response maintenance, by contrast, has connotations of poorly qualified personnel merely reacting to circumstances, being in essence not in control. This need not be the case; there is much to be said for getting the most out of each component before a timely and professional intervention is made, appropriate to the circumstances and meeting technical needs while simultaneously satisfying the user.

Unsustainability of PPM

The 'think big' approach of PPM is no longer fashionable; today's mantras are 'small is beautiful' and 'think globally; act locally'. 'Big government' and the 'nanny state' have been discredited. They gave rise to a so-called dependency culture in which individuals become habituated to having someone else to provide and do things for them and having a 'them' to blame. This has resulted in overprovision for many who may otherwise be satisfied with less; alternatively, cumbersome and maligned 'means test' mechanisms may be put in place, with associated administration costs. Either way, this takes money from front-line expenditure on repair works and puts it into spending where it is not critical or into administration.

For example, well-intentioned installations of full-house central heating may involve upheaval to furniture and floors and microbore piping chased into walls that then need redecoration or run within trunking hopefully placed unobtrusively on the surface. Often these systems are controlled by mechanisms that old people find difficult to adjust and to adjust to. Frequently the systems are designed to provide room temperatures determined by professionals for thermal comfort, only to find that tenants cannot afford the running costs – a situation known as 'fuel poverty'. Fuel poverty has been a significant problem in some local authority estates, particularly in areas associated with the former 'smokestack' industries of coal, steel and shipbuilding. These areas suffered greatly in the recession of the 1980s and many families became reliant on Social Security payments. Sherwood District Council in Nottinghamshire took a radical new approach. Instead of the usual 21°C, the design temperature was reduced to 16°C, with consequent reductions in not only radiator and boiler sizes and installation costs but also running costs.

By contrast, some tenants, habituated to the local authority's constant underfunding of maintenance and deferral of improvement programmes, have undertaken improvements to their home themselves. These home improvements may include refitting of their kitchens or bathrooms or installation of additional electrical sockets and occasionally minor structural works such as knocking two rooms into one. When the local authority later finds funding for the long hoped-for improvement programme or when the tenant moves on, it is then faced with difficult decisions about stripping out the tenant's own fittings in order to accommodate the 'standard' arrangement. Councils have an uncomfortable choice between, on the one hand, apparent insensitivity to tenants' care for and improvement of the property and on the other hand, a desire for simplicity and efficiency of maintenance of standard provision.

Both examples demonstrate problems related to the scale associated with PPM programmes. More sensitivity is required to individual wants and desires. A programme set up to provide its intended beneficiaries with a product that they neither want nor need, and that makes poor use of professional and material resources is not sustainable.

Need for large, non-local contractors

A concomitant of large-scale PPM programmes has been the tendency to let large contracts to large construction contractors. The underlying logic is that only the large, usually nationally operating contractors have the labour, project management and co-ordination skills required. Local authority staff or consultants called in to organise and undertake the stock condition survey and prepare the programmes and the associated tender and contract documentation are not generally willing to entrust such a large quantity of work to an assortment of small builders. These builders may be unreliable, poorly organised and unaccustomed to working to proper specifications. Professional surveyors' fees may be larger pro rata on smaller contracts and that will mean larger professional fees in total but there will also be more work involved; more risks, more headaches. Herein lies a paradox. Although surveyors recommend, and local authorities commission, large, national contractors to carry out PPM contracts, almost certainly the contractor will be subcontracting virtually all the work to the same small contractors that were thought inadequate.

Typically, PPM contracts may be worth many millions of pounds, dollars or euros over a period. For instance, a programme spending an average £1000 per dwelling over 5000 dwellings (a typical stock for a medium-sized largely rural local authority or housing association in the UK) represents a total expenditure of £25 million over a five-year period. This is the equivalent size and value of quite a substantial new-build project in the UK. On this basis, and especially for a client coming relatively new to such an approach and maybe new to projects of such magnitude, there is understandable attraction to the appointment of a large contractor with experience of management of a large project. When to this is added the complexity of organising a wide range of building tradespeople across a geographically dispersed 'site', the selection of a large, regional or national contractor becomes almost inevitable.

Often, prior to the adoption of PPM, owners or managers of property stocks will have built up a series of contacts with small local contractors able to respond to intermittent repair and maintenance needs. Usually these contractors, many of whom will be small and medium-sized enterprises (SMEs) of perhaps less than six employees, do not have the management or administrative infrastructures to be able to deal with even a part of a large PPM programme. Thus much experience and potential are lost.

The vast majority of building firms are very small. For instance, in the UK, 'in the region of 80% employ fewer than ten persons whilst less than 1% employ more than 250 personnel, (Chanter & Swallow, 1996), and in Australia the respective figures are very similar at 86% and 0.62% (de Valence, 2001). It is also the case that much of the building stock subject to PPM programmes, for instance local authority estates and other publicly owned stock, is situated in areas of social deprivation and unemployment. Sometimes authorities have employment programme priorities that encourage or require employment and maybe associated training of local unemployed people. There may be legal and other constraints around 'positive discrimination' in some countries. An 'incoming' contractor may have existing staff to 'import' or may be looking for local staff. In the UK, regulations apply to the situation where a public body ceases to undertake a service and decides to outsource the work. The Transfer of Undertakings (Protection of Employment) Regulations (TUPE, 1981) implemented the EC Acquired Rights Directive. Thus, sometimes 'in-house' staff or small contractors who previously worked for the client find themselves working for the new contractor, albeit perhaps only temporarily. Implicitly, PPM programmes restructure the local construction economy.

For some programmes supported by government funding, it is a condition of the award that a certain proportion of local labour be recruited. This may be thought to suggest that any unemployed and unskilled person can be recruited to construction and put to work straight away; that is to seriously underestimate the needs of the industry and its customers.

There are advantages to employing local labour. Local personnel will know their way around the area; they will live locally and know local suppliers. Thus transport and accommodation costs will be reduced and time saved. People who will remain in the area when works are completed may be thought to take more care to do a good job, to live more at ease with their neighbours and 'client', and hoping for the next piece of work too. They are more likely to stay with the job providing valuable continuity, avoiding recurrent recruitment and selection costs. This also means that local people learn what the client, as represented by the council or by the surveyor, expects by way of standards and procedures. This is reuseable and useful year after year as rapport and trust are built up. The value of trust to clients and contractors is being increasingly recognised and promoted through strategies such as 'partnering' and 'strategic alliances' (Latham, 1994; Bennett & Jayes, 1995; Wood & Smyth, 1996; Smyth & Thomson, 1999). More will be said on the building of and value of relationships and trust in Chapters 3 and 4.

The building up of skills, experience and confidence in the local building industry may be facilitated by the letting of PPM programmes initially to large contractors who import with them the necessary organisational and management expertise which is then 'trickled down' to the smaller firms. This may be an effective way of building up local capability for the future.

Adversarial nature of contracts

It would be virtually impossible in the UK to commission the implementation of a PPM programme without a formal contract. Although technically a contract only requires an offer, its acceptance and a 'consideration', it is common to prepare and sign documentation many pages long. Such contract documentation sets out clearly and for the avoidance of doubt the works to be carried out and the obligations of the parties to the contract. It is normally recommended that standard forms of contract be used; they have stood the test of time including periodic revision following clarification of points found to be unclear and contested in the courts.

Many small contractors are unfamiliar with and wary of even standard forms of contract. Much work is 'contracted' by exchange of letters or by word of-mouth, even though, as the movie mogul Samuel Goldwyn is reputed to have said, 'A verbal contract isn't worth the paper it is written on' (Johnston, 1937). A small contractor may find a 'Minor Works' form of contract satisfactory, but balk at a more weighty document. In the 1960s, the Royal Institute of British Architects published what became known as the 'RIBA form of contract'.

> '[The Form of Agreement for Minor Building Works] ... is a four-page document of thirteen clauses, being a very abbreviated form of the standard edition, intended for work which is not technically unusual or complex. It is especially suitable for contracts to carry out maintenance on existing buildings...' (Willis & George, 1970)

Over subsequent years the 'standard form' was developed into the JCT (Joint Contracts Tribunal) series of standard forms of contract. This series includes the 'normal' form (known as JCT63, thence JCT80, then JCT2000), the Minor Works form and subsequent forms, including an Intermediate form (IFC84) and a Measured Term Contract (MTC).

The MTC provides a vehicle for the letting of PPM contracts. In essence, a contractor undertakes to complete works of specified kinds within defined properties over a fixed period, at agreed rates per item. Typically, competitive tenders will be sought based on a Schedule of Rates (and perhaps approximate quantities) issued by or on behalf of the client in much the same way as a Bill of Quantities and/or a Specification by a quantity surveyor or architect for 'new-build' works. In the UK there are at least two major providers of standard Schedules of Rates, against which tenderers quote a variance, e.g. +2.6% overall or +13.1% on all works scheduled under 'plumbing' and −1.8% on all listed as 'carpentry', reflecting respective market conditions for instance. Whilst the MTC does give a degree of certainty to both client and contractor in terms of financial commitment and expectation, there is no guarantee of quantity of work, so it can be difficult to plan work and staff accordingly.

There can be further difficulties related to contracts. Whilst they exist for the avoidance of doubt, the importance of exactitude of each contract clause and the proliferation of increasing wordage to define meanings mean that many hundreds of thousands or millions of dollars may be at stake in a difference of interpretation of a single word or clause. The profitability, or even continuing existence, of a contractor or client may be in the balance. Often, therefore, many hours of expensive professional, lawyers' and court time will be spent on resolving disputed contract-related matters. Although procedures for arbitration, adjudication and alternative dispute resolution have been prepared, including in the UK their incorporation into legislation (Housing Grants, Construction and Regeneration Act 1996), it is far from certain that these address the fundamental causes of disputes.

The Latham Report (1994) identified fragmentation of the construction 'team' as a significant factor in the poor performance of the construction industry and consequent poor service to construction clients. Similar conclusions and recipes for change have been promulgated down the years: to name a few, Simon (1944), Emmerson (1962), Banwell (1964), NEDO (1964, 1983) and Wood (1975). Contracts represent and reflect differing concerns of the parties, their representatives and agents. There is not necessarily a commonality of objectives. Contracts may be used more to assist in the resolution of disputes after they have arisen, and in the hope of avoiding expensive litigation, than in order to do a good job, to get it right, and right first time.

Egan

Following on from his fellow knight Sir Michael Latham, Sir John Egan and his task force produced the report 'Rethinking Construction' in 1998. Egan reviewed the organisation and procurement of construction and the processes and products of the construction industry. Amongst his recommendations is a suggestion that consideration be given to the commissioning of work without formal contracts. Whilst this may seem heretical, and anathema to the legal profession, the recommendation is based on successful practice in Japan. The prerequisites are trust between all the participants and fair returns associated with building long-term relationships or 'partnering'. Smyth & Thompson (1999) have investigated the necessary conditions of trust for partnering.

In some ways consistent with Egan, PPM permits the construction of a substantial long-term relationship. Arguably, money could be saved by the elimination of administration associated with payments based on Schedules of Rates, if sufficient trust could be developed. The UK government has been encouraging the procurement of public building works based on 'best value' rather than on the basis of lowest tender. The UK's housing associations 'paymaster', the Housing Corporation, is requiring new housing and refurbishment to be procured on 'Egan' principles such as partnering.

Where PPM tends to fall short in relation to Egan is its reliance on paperwork or its modern electronic equivalent. Much energy and time are invested in undertaking condition surveys, documenting them, preparing detailed schedules and specifications, tender and contract documents, planning, organisation and management of the work, doing the work, checking that it has been done in accordance with the contract specifications and securing its rectification if necessary, administration of certification and payments, ... and so on. Perhaps an arrangement could be developed whereby the focus moves to service and satisfaction of building users within agreed limits of time and cost, with the final 'price' determined by reference to user-defined criteria – the ultimate in performance-related remuneration.

Millions of pounds and dollars are spent every year on construction litigation. I hazard to suggest, heretically, that much of the disputation in the construction industry arises *because* of the contractual arrangements intended to assist in their resolution. Many works have been dedicated to construction disputes and their resolution, e.g. Powell-Smith (1990), Latham (1994), Chappell (1997), Knowles (2000). The Housing Grants, Construction and Regeneration Act was introduced by Tony Baldry MP in 1996 specifically to assist the construction industry in resolving disputes, by setting up default procedures for adjudication and arbitration as alternatives to going to court.

By contrast, Egan suggested that significant construction works could be successfully carried out without a written contract, as Nissan UK has with its 130 principal suppliers. This would represent a significant 'leap of faith' by construction contractors and clients, and their advisors, who have grown up with an expectation of adversarial attitudes and consequential conflict. Either a lot of expense could be saved or scarce funds actually spent on effective maintenance, rather than talking about it! Egan estimated that implementation of more effective means of procuring construction work could reduce costs by as much as 30%. Procurement of maintenance services is discussed in more detail in Chapter 4.

Postmodern maintenance

The second half of the 20th century has seen the promulgation and deep penetration of the planned approach into the building maintenance industry. In part, this has been a response to increasing backlogs of maintenance as buildings have been starved of necessary investment and resources have been directed toward more attractive new-build projects. Planned maintenance programmes have been facilitated by the introduction of computerised project planning and management systems. However, many of these are in essence big number-crunchers that organise programmes of similar or, better still, identical work on the basis of economies of scale and predetermined priorities. The pound or dollar rules and the user, the ultimate 'client', has to get in line and wait his or her ordained place in the programme.

In the meantime advances in technology and management techniques have both enabled and required maintenance to be reconfigured in a more responsive mode, reflecting much more closely individual client needs and wants. By contrast with the previous centralised, planned approach, with a focus on the common good, this is a more 'postmodern' approach focused on the individual.

Summary

This chapter has discussed the provenance and persuasive logic of the 'think big' approach to maintenance that gave rise to the prevalence and pre-eminence of the planned preventive maintenance programme. The chapters that follow outline some alternative approaches that reflect more responsive and perhaps responsible attitudes and are aimed at getting closer to the customer. The influences of new management thinking, particularly from Japan, and rapidly changing telecommunications technologies and infrastructures such as the call centre have given rise to a whole new range of approaches, altogether 'postmodern'. This more considerate approach leads to the customer-focused concept of building care and represents a paradigm shift in building maintenance.

References

Almeida, A.T. & Bohiris, A.T. (1995) Decision theory in maintenance decision making. *Journal of Quality in Maintenance Engineering* **1** *(1), 39–45.*

Armstrong, M. (1994) *How to Be an Even Better Manager,* (4th edn.) Kogan Page, London.

Audit Commission (1986a) *Managing the Crisis in Council Housing.* HMSO, London.

Audit Commission (1986b) *Improving Council House Maintenance.* HMSO, London.

Audit Commission (2001) *Housing Inspection Service Annual Review: In Pursuit of Excellence.* Audit Commission Publications, Wetherby, Yorks.

Azar, K. (2000) The history of power dissipation. *Electronics Cooling* **6** (1), 42–50.

Banwell, H. (1964) *Report of the Committee on the Placing and Management of Contracts for Building and Civil Engineering Work.* HMSO, London.

Bennett, J. & Jayes, S. (1995) *Trusting the Team: The Best Practice Guide to Partnering in Construction.* Centre for Strategic Studies in Construction, University of Reading.

Berry, S. (2000) Maintenance of PVCu double glazing. Choices for local authorities. Unpublished MSc thesis, Oxford Brookes University.

BS 3811: 1984 *Glossary of Maintenance Management Terms used in Terotechnology.* HMSO, London.

BS 8210: 1986 *British Standard Guide to Building Maintenance Management.* British Standards Institution, London.

BS 4778, Part 3, Section 3.2: 1991 *Quality Vocabulary: Availability, Reliability and Maintainability Terms.* British Standards Institution, Milton Keynes.

BS 3811: 1993 *Glossary of Terms used in Terotechnology.* British Standards Institution, Milton Keynes.

Burke, E. (1791) Letter to a Member of the National Assembly, p.12.

Bushell, R.J. (1979) Preventing the problem. *Institute of Building Information Paper 11.* CIOB, Ascot.

Chanter, B. & Swallow, P. (1996) *Building Maintenance Management.* Blackwell Science, Oxford.

Chappell, D. (1997) *Powell-Smith & Sims' Building Contract Claims,* 3rd edn. Blackwell Science, Oxford.

Chartered Institute of Public Finance and Accountancy (1983) *Housing Management and Maintenance Statistics 1982–83 Actuals.* CIPFA, London.

Connor, R.A. & Davidson, J.P. (1985) *Marketing your Consulting and Professional Services.* John Wiley, Chichester.

Cox, D.R. & Oakes, D. (1984) *Analysis of Survival Data.* Chapman & Hall, London.

de Valence, G. (2001) Defining an industry: What is the size and scope of the Australian building and construction industry? *Australian Journal of Construction Economics and Building* **1** (1), 53–65.

Donnelly, J.H. & George, W.R. (eds.) (1981) *Marketing of Services.* American Marketing Association, Chicago, Illinois.

Drucker, P. (1967) *The Effective Executive.* Heinemann, London.

Egan, J. (1998) *Rethinking Construction: Report of the Construction Task Force.* Department of the Environment, Transport and the Regions, London.

Emmerson, H. (1962) *Survey of Problems before the Construction Industries.* HMSO, London.

Finch, E. (1988) *The Use of Multi-attribute Utility Theory in Facilities Management.* Proceedings of Conseil Internationale du Batiment Working Commission W70, Whole Life Asset Management, Heriot-Watt University, Edinburgh.

Hollis, M. & Bright, K. (1999) Surveying the surveyors. *Structural Survey* **17** (2), 65–73.

Housing Services Action Group (HSAG) (1980) *Organising an Effective Repairs and Maintenance Service.* Department of the Environment, London.

Hoxley, M. (1995) How do clients select a surveyor? *Structural Survey* **13** (2), 6–12.

Jardine, A.K.S., Banjevic, D. & Makis, V. (1997) Optimal replacement policy and the structure of software for condition-based maintenance. *Journal of Quality in Maintenance Engineering* **3** (2), 109–119.

Johnston, A. (1937) *The Great Goldwyn*, ch. 1. Ayer, New York.

Jones, K. & Collis, S. (1996) Computerised maintenance management systems. *Facilities* **14** (4), 33–36.

Knowles, R. (2000) *One Hundred Contractual Issues and Their Solutions.* Blackwell Science, Oxford.

Kobbacy, K.A.H., Proudlove, N.C. & Harper, M.A. (1995) Towards an intelligent maintenance optimisation system. *Journal of the Operational Research Society* **46**, 831–853.

Latham, M. (1994) *Constructing the Team.* HMSO, London.

Layzell, J.P.& Ledbetter, S.R. (1997) *Feasibility of Failure Mode and Effects Analysis in the Cladding Industry. Proceedings of Conseil Internationale du Batiment Working Commission W92 Symposium on Procurement: A Key to Innovation*, University of Montreal, May, pp. 375–383.

Lomas, D.W. (1997) Team inspections of high-rise buildings in Hong Kong and the UK. *Structural Survey* **15** (4), 162–165.

Moubray, J. (1990) *Reliability Centred Maintenance.* Butterworth-Heinemann, Oxford.

NEDO (National Economic Development Office) (1964) *The Construction Industry.* HMSO, London.

NEDO (1983) *Faster Building for Industry.* HMSO, London.

Oxley, R. (1998) More pitfalls for the surveyor. *Structural Survey* **16** (1), 5–6.

Pettit, R. (1983) *Computers Aid in the Management of Housing Maintenance.* Proceedings of BMCIS & BRE Seminar, Feedback of Housing Maintenance, pp. 35-40. Building Research Establishment, Watford.

Pitt, T. J. (1997) Data requirements for the prioritisation of building maintenance. *Facilities* **15** (3/4), 97–104.
Powell-Smith, V. (1990) *Problems in Construction Claims.* Blackwell Science, Oxford.
Saaty, T.L. (1980) *The Analytic Hierarchy Process.* McGraw-Hill, New York.
Scottish Local Authorities Special Housing Group (SLASH) (1979) *The Scottish Special Housing Association (SSHA) Planned Maintenance System.* SLASH, Edinburgh.
Shen, Q. (1997) A comparative study of priority setting methods for planned maintenance of public buildings. *Facilities* **15** (12/13), 331–339.
Simon, E. (1944) *The Placing and Management of Building Contracts: Report of the Central Council for Works and Buildings.* HMSO, London.
Smyth, H.J. & Thompson, N.J. (1999) *Partnering and Conditions of Trust.* Joint Symposium of Conseil Internationale du Batiment Working Commissions W55 and W65, University of Cape Town (proceedings on CD-ROM).
Spedding, A., Holmes, R. & Shen, Q. (1994) *Prioritising Major Items of Maintenance in Large Organisations.* Proceedings of RICS Research Conference, Salford University, December, pp. 123–131.
Then, D.S.S. (1995) Computer-aided building condition survey. *Facilities* **13** (7), 23–27.
Triantaphyllou, E., Kovalerchuk, B., Mann, L. & Knapp, G.M. (1997) Determining the most important criteria in maintenance decision making. *Journal of Quality in Maintenance Engineering* **3** (1), 16–28.
Tsang, A.H.C. (1995) Condition-based maintenance: tools and decision making. *Journal of Quality in Maintenance Engineering* **1** (3), 3–17.
TUPE (1981) *Transfer of Undertakings (Protection of Employment) Regulations.* Statutory Instrument SI 1981/1794. HMSO, London.
Wheiler, K. (1987) Referrals between professional services providers. *Industrial Marketing Management* **16**, 191–200.
Willis, A.J. & George, W.N.B. (1970) *The Architect in Practice*, 4th edn. Crosby Lockwood, London, p.70.
Wilson, A. (1984) *Practice Development for Professional Firms.* McGraw-Hill, Maidenhead.
Wittreich, W.J. (1966) How to buy/sell professional services. *Harvard Business Review*, **44**(2), 127–138. Cambridge, Massachusetts, March/April 1966.
Wood, B.R. (1986) Surveying the estates. *Housing* **22**(2), 24–25.
Wood, B.R. & Smyth, H.J. (1996) *Construction Market Entry and Development: The Case of Just in Time Maintenance.* Proceedings of 1st National Construction Marketing Conference, Oxford Brookes University, pp. 17–23.
Wood, K.B. (1975) *The Public Client and the Construction Industries.* NEDO, London.
Zeithaml, V.A. (1981) How consumer evaluation processes differ between goods and services. In: *Marketing of Services* (Donnelly, J.H. & George, W.R., eds). AMA, Chicago, Illinois.

3 Just in Time: Gurus from East and West

While many in the building maintenance 'industry' were developing extensive planned and so-called preventive maintenance programmes, the business community more generally had been falling under the global influence of waves of 'management gurus', each with their own prescription for business success. This chapter examines how the concept of 'just-in-time' delivery, developed in the USA and applied successfully in industry in Japan, was seen to have application to building maintenance. It is based on a funded study in UK, which identified innovative practice that the author christened 'just-in-time maintenance'. Subsequent chapters investigate the penetration of further business-related processes and priorities.

Gurus

Armstrong (1994) and Norton & Smith (1998) give good overviews of the growth and penetration of the management guru. Box 3.1 gives a timeline of the genesis of management prescription.

Box 3.1 The evolution of management publications.

Year	Author	Title
1911	Taylor	Principles of Scientific Management
1916	Fayol	General and Industrial Management
1924	Weber	The Theory of Social and Economic Organisation
1933	Mayo	The Human Problems of an Industrial Civilisation
1954	Drucker	The Practice of Management
1954	Maslow	The Hierarchy of Needs
1959	Herzberg	The Motivation to Work
1960	McGregor	The Human Side of Enterprise
1960	Levitt	Marketing Myopia
1961	Likert	New Patterns of Management
1962	Chandler	Strategy and Structure
1964	Blake & Mouton	The Managerial Grid
1965	Ansoff	Corporate Strategy
1969	Drucker	The Age of Discontinuity
1973	Adair	Action-Centered Leadership
1976	Handy	Understanding Organisations
1978	Argyris & Schon	Organisational Learning
1979	Crosby	Quality is Free
1980	Belbin	Management Teams: Why They Succeed or Fail

1980	Hofstede	Culture's Consequences
1980	Porter	Competitive Strategy
1980	Revans	Action Learning: New Techniques for Management
1981	Pascale & Athos	The Art of Japanese Management
1982	Honey & Mumford	Manual of Learning Styles
1982	Ohmae	The Mind of the Strategist
1982	Peters	In Search of Excellence
1983	Kanter	The Change Masters
1983	Levitt	Marketing Imagination
1984	Kolb	Experiential Learning
1985	Ohmae	Triad Power
1985	Porter	Competitive Advantage
1985	Schein	Organisational Culture and Leadership
1986	Deming	Out of the Crisis
1988	Juran	Juran on Planning for Quality
1989	Bennis	On Becoming a Leader
1989	Covey	The Seven Habits of Highly Effective Managers
1989	Handy	The Age of Unreason
1989	Kanter	When Giants Learn to Dance
1989	Mintzberg	Mintzberg on Management
1989	Peters	Thriving on Chaos
1990	Adair	Understanding Motivation
1990	Hamel & Prahalad	The Core Competence of the Corporation
1990	Hammer	Re-engineering Work: Don't Automate, Obliterate
1990	Kotter	A Force for Change
1990	Pascale	Managing on the Edge
1990	Schien	Career Anchors; Discovering Your Real Values
1990	Senge	The Fifth Discipline: ... the Learning Organisation
1992	Ohmae	The Borderless World
1992	Peters	Liberation Management
1993	Hammer & Champy	Re-engineering the Corporation
1993	Morgan	Imaginisation: The Art of Creative Management
1993	Trompenaars	Riding the Waves of Culture
1994	Hamel & Prahalad	Competing for the Future
1994	Handy	The Empty Raincoat
1994	Mintzberg	The Rise and Fall of Strategic Planning
1995	Barnatt	Cyberbusiness: Mindsets for a Wired Age
1995	Nonaka & Takeuchi	The Knowledge-Creating Company
1995	Tulgan	Managing Generation X
1996	Kanter	Becoming World Class
1996	Kaplan & Norton	The Balanced Scorecard
1997	De Geus	The Living Company
1997	Handy	The Hungry Spirit

The galaxy of gurus, the procession of purveyors of prescriptions for business success, is endless. With a succession of such recipes it is easy to become cynical and to consider management theories to be no more than passing fads. That would be to marginalise the value of thinking about business and its organisation. A fundamental underpinning of this book is the contention that buildings and their occupants would benefit from more thought being given to their management and to the selection of appropriate policies and procedures. It is not intended, however, to give detailed analyses of the application to building management or maintenance of even a fraction of the theories represented by the gurus above. Arguably, one would have expected that to be happening as a matter of course within enterprises worth many millions of pounds or dollars. The comparative sparseness of the literature of building maintenance theory suggests that this has been little developed over a comparable period.

The PPM paradigm has held sway for a quarter of a century. The seminal works by Lee (1987) and Seeley (1976) describe PPM as maintenance 'organised and carried out with forethought and control', 'to a predetermined plan or prescribed criteria' and 'intended to reduce the likelihood of an item not meeting an acceptable condition'.

As indicated in Chapter 2, the application of PPM has been expounded by, for instance, Bushell (1979), the UK Housing Services Advisory Group (1980) and Audit Commission (1986, 2001), and exemplified in published examples by the Scottish Local Authorities Special Housing Group (1979) and Pettit (1983). The literature is not, however, without reservations about PPM.

Noble (1980) described the logic of PPM as 'a stitch in time saves nine'. He then observed that 'the ideal maintenance situation in which the condition of the property is kept within predetermined limits by a pre-planned programme of preventive work is never achieved in practice and is in fact unattainable except at impossibly high cost'. He further argued that:

> 'Some industries, e.g. commercial aviation, have maintenance services which more nearly approach the ideal because the revenues lost during downtime and the consequences of failure are such as to make reliability of prime importance: high maintenance costs are therefore acceptable. In buildings, defects are more readily tolerated.'

Lee himself described PPM as 'a concept which is probably more applicable to plant and equipment'. Bushell (1979) suggested that PPM is worthwhile if:

(1) it is cost effective
(2) it is needed to meet statutory requirements
(3) it meets clients' operational needs
(4) it will save running maintenance
(5) there is work for the craftsman rather than pure inspection.

The author had much experience of the development and implementation of maintenance and improvement programmes in the UK in the 1980s, principally in relation to substantial local authority housing stocks. When entering university service in the 1990s, the opportunity arose for the author to 'revisit' PPM. Together with a colleague, a small research project was proposed and funding obtained. Work was undertaken to investigate the penetration of PPM in areas other than publicly maintained housing. An outcome of that research was the identification of 'the abandonment of preplanned maintenance in favour of just-in-time maintenance' (Smyth & Wood, 1995).

Japan

The American management guru W. Edwards Deming was very influential in Japan. His approach to quality was key to the implementation of 'just in time'. The loss of the buffer of long lead times meant that there was little or no time in which to correct defective work. Deming put the emphasis on avoidance of defects rather than on their detection and correction. The responsibility for quality rests with the person doing the work, not with an inspector. It is a common failing in the West to assume that workers take little pride in their work, obeying McGregor's 'Theory X'. Often their tasks have been deskilled and 'those who perform them are supposed to be motivated solely by factors such as piecework and bonus schemes' (Wheatley, 1992). On the contrary, people do not like producing defects and they take satisfaction in improving performance. Indeed, operatives are key in identifying problems and ways in which defects may be reduced or eliminated. The work of Atkinson (1998) in the UK construction industry testifies to the importance of the human element in defects and their avoidance.

To understand how 'Just in time' came to prominence in Japan, it is important to appreciate that despite the increasing promotion of the idea of a single 'global market', there are significant cultural differences around the world. In business, as with relationships generally, it is dangerous to be insensitive to these. Geertz (1973), Hofstede (1980) and Trompenaars & Hampden-Turner (1997) have written authoritatively on culture. A useful review of culture and management specifically in Japan is provided by Hayashi (1988).

Hayashi offers observations on a number of aspects of life and work in Japan and particularly in relation to ways of working, values and time. Two quotes from Hayashi are offered here to give a flavour.

> 'In Japan the way something is done, not just the result, is much admired. In judo and kendo the performer must master many stylised moves. The formal tea ceremony is another example.'

> 'In the West, time is perceived as an arrow moving rapidly toward a target. Japanese see time unfolding as an endless scroll, without start or end, and the present advances as the scroll uncurls.'

At Toyota, a production engineer called Taiichi Ohno introduced a system of shorter production and supply lines, known as 'kanban'. In contrast to the normal supply-led (or 'push') systems, operations were demand led ('pull'). Kanban is a Japanese word meaning 'ticket' or 'sign'. The kanban would represent the 'works order' passed to the supplier when the next batch of a particular part was required, thus avoiding stockrooms. Ohno's approach concentrated on elimination of waste. According to Wheatley (1992), 'not surprisingly, he [Mr Ohno] is sometimes called "the father of Just in Time"'.

Just in time

> 'The first thing to be said about JIT is that it is most definitely not a software package; it is a philosophy ... a philosophy of common sense. The essence of this philosophy can be stated using two expressions which respectively sum up the positive and negative aspects of JIT: the "habit of improvement" and the "elimination of wasteful practices" ... an ongoing crusade by both workers and managers.' (Dear, 1988)

According to Riggs (1987):

> 'Advocates of JIT claim it is a revolutionary concept that all will have to adopt in order to remain competitive ... the basic approach is to continually reduce costs by stressing the elimination of waste: no rejects, no delays, no stockpiles, no queues, no idleness and no useless motion.'

In the manufacturing and retail sectors the main advantage from the supply end is to reduce stockpiles and hence working capital. From the demand side it is a customer-orientated approach. The customer or client makes demands for a product or service at short notice. Market needs are satisfied more quickly because the supply chain is cut in duration and cashflow is increased. It is a marketing tool as well. As goods or services are produced in response to demand, this gives the opportunity to customise the product or service to the specific requirements of a client or a group of customers. In manufacturing this has led to a growth in small batch production catering for niche rather than mass markets (Wilson, 1991).

Muhleman *et al.* (1992) argue that '...JIT demands perfect equipment maintenance, since breakdowns cannot be tolerated. It is not sufficient to speed up repairs to minimise downtime: breakdowns must be eliminated through an effective prevention strategy'. Therefore, 'just as in the achievement of quality... operators must be given the tools (and this means providing the appropriate training) to be able to detect, find and eliminate potential causes of trouble before they manifest themselves in a system failure'.

Dear (1988) provides a good review of issues in working towards JIT. He identifies the value of a gradual approach to improvement, starting from a base of knowledge of how things are done at present, and counsels against the dangers of all-singing, all-dancing complexity. The 'KISS' principle (Keep It Simple, Stupid) has much to commend it. There is a need generally to improve data and forecasting based upon it. There is no value, however, in data for its own sake; there can be too much. This is borne out by a small research study by the author. A well-known computer company was amongst a sample investigated in relation to 'intelligent building controls'. The company had a sophisticated building and energy management system. What was found was that although significant quantities of data were automatically collected – for example, energy inputs, external and internal temperatures, humidities, ventilation rates, lights on/off, power utilisation – there was no energy manager. There was no-one with time and responsibility to analyse the wealth of data and turn it into useful information; hence no identification of problems and possible improvements. This was despite the likelihood that cost savings generated would probably repay costs several times over, producing significant year-on-year improvements.

Dear recommends that attention be given to improving forecasting. Trends, seasonality, cycles, random events should be identified and put together with market knowledge, and maybe gut feeling, to plot possible futures. Graphical representations can be very helpful to see 'patterns'. The author's experience in relation to building stock condition surveys bears this out. By plotting on an estate plan the condition of chimney-stacks, it was possible to detect 'hotspots' of poorer condition. Conversations with residents confirmed that those had been built by a different contractor from the one used on the remainder of the estate.

Interpretation and extrapolation of data can be difficult. There is a danger that forecasters will be unduly influenced by others – either overly optimistic or pessimistic. A good way of proceeding against a background of uncertainty may be to set up a pilot study. However, as Dear identifies, it is not usually difficult to identify wasteful practices, the problem is in managing to do something about them. We tend to prefer to talk about problems than to solve them. He contends that we are good at reacting to crises because we get a lot of practice at that.

Hewlett-Packard, Japan and quality

Through the late 1970s and the 1980s American and European manufacturing businesses became increasingly exposed to severe competition from Japanese companies. Whole industries (e.g. TV, VCR) were taken over or dominated by the Japanese.

Cole (1999) assesses systems applied at Hewlett-Packard in learning from Japanese experience through a joint venture, Yokogawa Hewlett-Packard (YHP). A stunning example of accomplishment is given. The 'wave solder rate of non-conformity' was reduced from 4000 parts per million (ppm) in 1977 to 40 ppm in 1979 and eventually, with the help of quality circles, to 3 ppm in 1982. This was attributed to the operationalisation of six principles of total quality control (TQC), as identified by Mozer (1984).

(1) A commitment to continuous quality improvement led by top management.
(2) The collection of data in analysing problems (management by fact).
(3) Clarification of who was responsible for action in daily work and in problem solving.
(4) Systematic gathering of feedback from internal and external customers.

(5) Use of the Deming cycle (Plan, Do, Check, Act) as a problem-solving process to achieve permanent solutions.
(6) Use of statistics as a management tool.

In seeking to understand its loss of market to Japanese competition, Hewlett-Packard examined their processes and controls. 'While HP still enjoyed a reputation for high quality, it discovered that some 25% of its manufacturing costs were accounted for by the costs of fixing quality problems' (Young, 1983). As a response, HP's CEO David Young instigated a quality initiative dubbed as the '10× initiative', which called for a tenfold improvement in hardware quality (as measured by reliability) over the rest of the decade. By 1988 the quality of semiconductors equalled that of the Japanese. Defect rates had been reduced to about one part in 20 000, a 40-fold improvement from 1980 (Main, 1994). Throughout the 1990s Hewlett-Packard was consistently ranked first in customer satisfaction and reliability surveys (Ristelheuber, 1994; American Customer Satisfaction Survey, 1995-7; Hewlett-Packard, 1997).

Zen and kaizen

Japanese practices promoted and popularised process improvement. The quality movement brought the concept of continuous improvement (kaizen), to which the business guru Tom Peters added customer focus. In their seminal work *In Search of Excellence* (1982), Peters & Waterman compared the products of the American motor industry with that of the Japanese. They concluded that 'lack of care and attention is a detrimental aspect of the finish given by so many companies to the work they undertake'.

What, exactly, is the nature of Japan's magic?

> 'They excel in the quality of fits and finishes, moldings that match, doors that don't sag, materials that look good and wear well, flawless paint jobs. Most important of all, Japanese cars have earned a reputation for reliability, borne out by the generally lower warranty claims they experience. Technically, most Japanese cars are fairly ordinary.' (Burck, 1980, quoted in Peters & Waterman, 1982)

Peters & Waterman quote from Pirsig's *Zen and the Art of Motorcycle Maintenance* (1974): 'While at work I was thinking about this lack of care in the ... manuals I was editing ... they were full of errors, ambiguities, omissions and information so completely screwed up [that] you had to read them six times to make any sense out of them'. They also quote Professor William Abernathy (Gall, 1981): 'The Japanese seem to have a tremendous cost advantage ... The big surprise to me was to find out that it's not automation ... They have developed a "people" approach'. This was further exemplified by an observation from Kenichi Ohmae (1981), head of McKinsey's Tokyo office, that 'in Japan *organisation* and *people* are synonymous'.

Quality, reliability and defects

Cole (1999) tracks the rise in interest in quality in a sequence of what he calls 'fads' over the latter part of the 20th century. He quotes Garvin (1988) in justifying a rationale for this interest as a growth in consumerism in the 1960s and resultant product recalls and liability suits.

In a previous text, Cole (1981) indicated that the Institute of Social Research at the University of Michigan reported that 'public concern about auto quality doubled between 1968 and 1975'. By the 1990s this had translated into a huge growth of the quality 'business', with Cole reporting (1999) that: 'By 1991, *Business Week*'s special issue on quality reported a Boston Consulting Group study that concluded American companies were paying out $750 million a year to 1500 third-party providers of advice and materials on quality improvement (Byrne, 1991)'.

The Japanese identified in the 1980s that Western manufacturing firms had underestimated the importance to customers of reliability. It is contended that the Japanese approach of continuous improvement (kaizen) contributed significantly to quality performance by eliminating waste and rework in business processes. The Americans, however, with a 'tendency to push the technological frontier, often fail to use the less advanced processes and products that can be mass-produced reliably' (Stalk & Hout, 1990).

Regarding defects, Cole quotes the work of Okimoto *et al.* (1984) in relation to semiconductors that showed that the best American companies produced almost six times the defect rate of the best Japanese, and the worst American performance was over 14 times worse than the worst Japanese performance. Finan & LaMond (1985, quoted by Cole, 1999) reported that the cumulative yield for 64k chips was 23% for American producers and 38% for Japanese producers.

Cole constructs a table of 'quality fads', from which the following has been derived.

Late 1970s/ early 1980s:
- quality circles
- competing gurus (Crosby, Deming, Juran)
- cross-functional teams
- quality function deployment

Mid-1980s:
- customer focus
- continuous improvement
- partnering
- benchmarking

Early 1990s:
- business process re-engineering
- ISO 9000

Cole argues that this sequence reflects roughly the pattern of the automotive industry; home appliances, electronics and the service sector followed and higher education and hospitals are recent arrivals. He summarises the contributions of the 'competing quality gurus' through the first half of the 1980s, refering to this as a period of 'quality by exhortation' and distilling their contributions thus.

- *Crosby* – costs of non-conformance to requirements; top-down control orientation using the language of management; focus on attitudinal change; employees to adopt a zero-defect mentality; little focus on customers
- *Deming* – statistical control; reducing sources of variation; fewer suppliers, and longer relationships (supply chain management); developing cycles of continuous improvement
- *Feigenbaum* – systems approach
- *Juran* – budgeting and project approach

Box 3.2 is drawn from Cole's synthesis of developing approaches to quality. He contrasts 'old' and 'new' quality models. The old model reflects the 'scientific management' model of Taylor, seeing quality in terms of meeting a standard, and fails to produce the best quality product. Increasingly the aim is not only to meet customer expectations but to exceed them.

Box 3.2 Old and new quality models.

Old quality model

Internal orientation
Conformity to standards
Cost reduction
Quality is a functional specialism
Quality not a competitive feature
Quality a function carried out by experts
Downstream focus on inspection
Detect defects and repair
Quality department
Departments pursue departmental goals

New quality model

Customer orientation
Market approach
Anticipate customer needs
Quality an umbrella theme
Quality as a competitive strategy
Quality involves all employees
Upstream prevention; build in quality
Continuous improvement
Training
Cross-functional co-operation to achieve company objectives

Over recent years there has been growth in interest in 'standard' quality systems. In Britain this was initially manifested through BS 5750, which has now been subsumed within the ISO 9000 series. These systems are commonly misunderstood. ISO 9000 certification does not measure the quality of a product or service – it only indicates that a company has fully documented its quality control procedures and confirms that they are working in adherence to them. Some customers expect to relate quality to registration, but this need not be so. ISO 9000 is based on a simple methodology – document what you do, do what you document, and verify that you are doing it. Quality system registration does relatively little to guarantee the quality of the product or service other than its consistency, which could be consistently poor.

Stock and stockrooms

For a manufacturer, work in progress (WIP) represents stock tied up and expenditure on materials and labour; for a customer it represents waiting time. With JIT an aim is to reduce or eliminate unnecessary WIP, such as that which occurs in the assembling of batches of similar work or long lead times between order and execution, and thereby to minimise waiting time for the customer. It also avoids the need for stockrooms that would otherwise hold not only stock awaiting despatch but also stock for which there may never be a customer, or perhaps one, 'once in a blue moon'. The storage process and related staffing and space are wasteful and ultimately unproductive 'on-costs'.

A corollary of JIT is greater attention to quality. Without the buffer of long lead times, it is important to avoid the costs and inconveniences of repeat visits to correct inadequate or defective work. The work of Deming is relevant here. The closer the relationship between the customer and the maintenance operative, the greater the concern on the part of the operative to 'get it right', to identify what the customer actually wants. It is also more likely that the job will be done well and promptly, and that the customer will be satisfied. Traditional approaches often assume that operatives are poorly skilled and motivated largely by incentives such as bonus schemes. JIT presupposes that people at all levels take pride in their work and take responsibility for quality.

Ohno's approach at Toyota concentrated on elimination of waste; this requires investigation of existing processes and identification of opportunities for improvement. However, as Wheatley identified (1992) in relation to production environments, '... team members may have had little experience in sitting around a table and formulating ideas, let alone in communicating them to others via a flipchart or blackboard'. The experience of maintenance staff may not be dissimilar to this. There is a need to promote the management skills of maintenance personnel. Atkinson (1998) identified that 'managerial influences underlie many errors leading to defects [and] ... as a consequence, the continual emphasis placed by technical publications on correct technical solutions ... is misplaced'.

Buyer power and supply chain management

According to Wheatley (1992), in contrast to the traditional way of working with suppliers, expending time and energy on 'haggling' about prices and delivery, with emphasis on discounts for quantity, JIT is based on building a longer term relationship. This chimes in with the observations of Latham (1994) and Egan (1998) in relation to the problems of fragmentation and opportunities for alliances and 'partnering' in the construction industry. It is suggested that the mutual dependency associated with single sourcing reduces risk. 'The traditional approach to purchasing was to regard suppliers as adversaries' (Wheatley, 1992). The focus would generally be on getting the lowest price and, by 'shopping around', securing discounts for quantity and placing orders with a multiplicity of suppliers. Time and effort are wasted in complaints about quality and late deliveries. JIT works on a basis of longer term relationships with a smaller number of suppliers treated as partners, co-operating to achieve improved performance. Benefits include lower stock levels, shorter lead times, better quality and lower costs, with less risk. Those in such a supply chain are able to work together consistently to identify ways to secure continuous improvement.

Both Latham and Egan stress the value of overcoming the antiproductive adversarial relationships that have characterised the construction industry. They propose more open and trusting dealings between client and contractor and so onward down through a more constant and consistent supply chain, speaking of 'partnering' (Bennett & Jayes, 1995), alliances and teamwork. Long-term relationships are rare; that between Bovis Lend Lease and Marks and Spencer is habitually referred to. Repeat business has historically been low in construction, at around 20% (Gronroos, 1990, 1994). The Nordic School of Marketing (referred to in Chapter 1) suggests that 'relationship marketing', by comparison with the traditional 'marketing mix' strategies of price, product, place and promotion, offers opportunities to enhance client satisfaction and increase repeat business and profitability. The influence and impact of this and other approaches have been considered in relation to the procurement and provision of building maintenance services through a series of studies that informs this and subsequent chapters.

Smyth & Wood, 1995

Building on the author's extensive experience in the development and implementation of PPM programmes, it was decided to carry out further research with the encouragement and assistance of Hedley Smyth, then a colleague at Oxford Brookes University, who brought a wealth of experience in the marketing of construction services. Our programme of research intended to investigate the different responses to procurement and maintenance across building types with different ownership, user and utilisation patterns. The proposed building types were retail food stores, self-build housing and international conference and convention facilities. The aim was to establish the degree of planned maintenance and effective response, identifying 'best practice' and the conditions for improved implementation according to the pattern of ownership and use.

Our pilot research study of the retail food sector, carried out in 1994–5, identified a set of practices and outcomes that challenged the 'conventional wisdom', with a demand for more proactivity in using cost-in-use criteria generally and in planned maintenance in particular. The research uncovered a polarised response to the management of the estate. Analysis showed this to be part of a 'back to core business' philosophy coupled with a restructuring of the workforce, including redundancy.

One result was that the retained staff, those directly employed, concentrated upon the specification of buildings and equipment. These cost and specification analyses are based upon performance specification. Capital cost and life costs are critical elements in determining quality to price ratios. These are used for procurement of building components and equipment, and in performance specification for the architect and engineer's brief for those elements that cannot be predetermined.

The second outcome was the abandonment of preplanned maintenance in favour of just-in-time maintenance through letting construction and maintenance contracts to subcontractors. The demanded response times were critical and linked to the payment of the subcontractor. Those entering this business were both the traditional contractors and contractors from a security and industrial cleaning background who were growing their 'facilities management' side.

The research identified the way in which these contractors operated in this new and growing market, the approaches adopted by the different players and the constraints under which they operated. The research also considered the benefits and the potential constraints of the just-in-time approach for the client, the retailer. The theoretical underpinnings originally came from the body of concepts for cost-in-use research which were then evaluated in relation to just-in-time concepts.

Background to the study

Work addressing the poorly regarded (and consequently poorly rewarded) activity of building maintenance over the last 10–15 years has focused significantly on two areas: lifecycle costing (LCC) and planned preventive maintenance (PPM). PPM was discussed in some detail in Chapter 2; LCC is discussed further in Chapter 11.

Feedback is an important component in informing repair and replacement decisions in both LCC and PPM. Skinner (1983) is a strong proponent of the benefits of feedback 'directed towards such topics as repair frequency and service life of components, preventive maintenance, workload forecasting and the effect of design/construction on maintenance cost levels'. He links this to 'computer-based origination of maintenance job instructions for a group of buildings'.

NBA Construction Consultants (1985) were commissioned by the Audit Commission to present evidence and recommendations on maintenance cycles and life expectancies of building components and materials.

There are advocates that buildings should be treated more like a car (RICS, 1990) or with the rigour of an aircraft (Noble, 1980). Checks, services and component replacement would be considered at the design stage and throughout the life of the building and across the whole estate. This is a long way from recent practice and for some good reasons.

First, many commercial buildings around the world are commissioned by developers who have minimal interest in their cost in use. One such developer, Land Securities, was responsible for pioneering in Britain the full repairing lease, whereby building procurement is totally separated from responsibility for maintenance. The latter is the sole province of the tenant, who bears not only the cost of maintaining the property for their own purposes but also maintaining the asset value of the building for the benefit of the developer (Marriott, 1967; Smyth, 1985).

Second, maintenance has an opportunity cost for the commercial occupier. The market pressures to declare high profit levels or invest in areas that bring quick and tangible returns are great and investment in building maintenance is less tangible. It is also an easy target for expenditure reductions during recessionary times. Third, most commercial companies are characterised as having little property expertise at key decision-making levels. In Britain it was estimated that only one in six organisations has property expertise at board or senior management level on which it could draw. Fourth, in the case of owner-occupiers the rate of moving can be faster than the consequences of delayed maintenance. In inflationary times and during a booming property market this is exaggerated as surveys that pick up defects are unlikely to result in lower prices. Many surveys also focus upon existing defects and defects that may emerge in the short term rather than the longer term. These are some of the main reasons for a lack of focus on planned maintenance and cost-in-use considerations at the design stage.

Returning to core business

However, there have been some signs of change. During the last decade of the 20th century there was a greater commercial awareness of the effect of maintenance upon the asset value of companies and hence gearing, ability to borrow and stock market rating. This was coupled with the sale of surplus land holdings. The business philosophy of returning to 'core' business led to a shedding of in-house expertise and a growing sophistication in outsourcing, which in construction terms is subcontracting. Advocates of planned maintenance could possibly see this as a development that may in the long term lead to their ideal being fulfilled, especially as the costs of repairs grow through years of neglect during recession.

The research suggested, however, that the 'ideal' of planned maintenance was unlikely to emerge under current and future circumstances. The analysis demonstrated a growing concern about cost in use at the procurement stage filtering through to all capital procurement decisions, including the design of buildings. The research also showed that planned maintenance was being reduced and discarded. The retail food sector was leading this change. Their response in Britain has been to:

- reduce overheads and the salary bill by reducing maintenance staff to a minimum
- return to core business
- subcontract maintenance work
- maximise the life of all capital equipment and buildings.

Yet their turnover is large and they have been amongst the most profitable businesses during recession coupled with large property investment programmes, particularly new out-of-town stores in Britain. Their business is highly sensitive to unexpected costs, especially where defects in store equipment and buildings would lead to customer dissatisfaction or even the closure of a store. The consequent loss of turnover and profit would be unacceptable. How had they managed this situation?

The analysis demonstrated that it was precisely the response of returning to core business during recessionary conditions that was generating new approaches to maintenance. The analysis showed a polarisation between the use of more stringent specification at the procurement stage, based upon performance over the lifecycle of the equipment or building, and the requirement to delay maintenance in order to maximise the use of all capital items including the buildings. Procurement was being based around value for money, perhaps more precisely expressed as a quality-to-price ratio. Maintenance was far from planned. It was being subcontracted to others who were then required to provide immediate response to any defect as it emerged. That analysis identified the birth of what has been christened 'just-in-time maintenance'.

Just-in-time maintenance

Just-in-time maintenance borrows its mode of operation from just-in-time manufacturing and delivery. JIT practices are well established in other industries (Droy, 1986; McLachlin, 1990; Duclos *et al.*, 1995) and there are instances of JIT being used in the maintenance of manufacturing equipment (Duclos & Spencer, 1994).

Advances in technology have facilitated the development of devices which reduce costs while improving operational performance such as computer-based systems which enable the detection of faults before severe difficulties occur, e.g. sensing devices to monitor factors such as machine vibration, local temperatures, pressures, consumption of lubricants. Changes in such factors tend to indicate changes in the condition of equipment and can give timely warning of approaching failure. Muhleman *et al.* (1992) call this approach 'predictive maintenance' and this is a prerequisite for JIT maintenance to function.

JIT practices are rare in the sphere of the built environment. They are particularly rare in contracting; however, there is evidence of them being applied in the Finnish construction sector, where off-site prefabrication is widely used (Pajakkala *et al.*, 1993). The Japanese have also been using a customer-orientated approach linked to prefabrication in housebuilding (Bottom *et al.*, 1994). It may be that the UK government's promulgation of programmes based on Egan (1994) principles will promote wider development of such possibilities.

It is the small batch approach to production and delivery plus the client orientation that is significant in the maintenance market. The small batch approach can be applied to maintenance where defects tend to be different each time because over a period of time data can be generated from experience about the frequency and likelihood of a particular building fault occurring. This enables the construction of the 'small batch' or, more accurately, the identification and anticipation of generic maintenance responses. The client-orientated approach means that the buyer, in this case the food retail operator, has the power in the marketplace to demand a type of response that reconfigures the market; that is, how the work is done. The power comes from their increasingly oligopolistic position in that sector and as a major client in the construction and facilities management industry. How has JIT maintenance emerged and risen to prominence among food retail operators?

Emergence of JIT maintenance

The growth of food retail operations in the 1980s, especially in out-of-town locations, has changed management and customer expectations. When sites were in traditional town centres, buildings tended to vary greatly in size and shape, with an overlaid corporate identity of signage and fittings. The business focus was largely on returns, resulting in tendencies to neglect maintenance, to make good the symptoms, for instance patch-repairing the prevalent flat roofs. This had a deleterious effect not only on the properties but also on their value as financial assets.

The unpredicted jump in interest rates in the early 1980s made the price of all new investments difficult to justify (Riggs, 1987). It also caused the whole concept of 'safety stocks' to be questioned. Efforts to reduce the amount of capital tied up in inventories in order to reduce interest charges and improve balance sheets, the increase of just-in-time deliveries and the reduction in margins from town centre stores were all significant in management reappraisal.

The newer, larger stores have tended to be characterised by standard layouts with minimal storage areas, well lit and highly serviced with large runs of refrigerated cabinets and with standard fittings including checkouts capable of recording and relaying copious information about sales. These developments have coincided with management theories that promote concentration on core business, re-engineering the organisation and outsourcing the non-core business. This has resulted in two, at first apparently paradoxical, changes of direction.

Organisational consequence

First, from a situation where maintenance was dealt with largely locally, *ad hoc*, and unsatisfactorily, this has moved to being regionally organised, with the backlog of fabric repairs brought under control. Responsibility for ordering repairs has been restored to the local level but using contractors organised regionally, generally on the basis of term contracts.

Second, the power of the 'buying function' has been extended from the retail products to the building infrastructure. The market researchers have long informed the layout and style of the stores such as the width of aisles, heights of displays, lighting quality. It has become common for the large retailers to design their own cabinets and have them made specially to their needs, their buying power bringing an economy of scale.

The electronic connection of the checkouts direct to headquarters means that information on how a particular line is selling is known almost instantly and decisions on buying, pricing, positioning and so on can be taken accordingly. These electronic links enable further information – about the building and its fittings – to be received centrally or transmitted remotely. It is thus possible for central management or a contractor to know that the temperature of a particular refrigerated cabinet at a particular store is rising. By combining this information with parameters derived from commonly collected and assessed performance data, it is possible to determine when an unacceptable situation may be reached so that remedial JIT maintenance action can be ordered. A precondition for the emergence of JIT maintenance was therefore the centralisation of control for this function.

Findings from the research

This study of the food retail sector found support for the possibility and incidence of just-in-time maintenance, whereby prescheduled work was increasingly being replaced by a call-out just sufficiently in advance or at the time of affecting the operations of the client. This has to be seen as a polarised response to greater care in capital equipment and building procurement and costs, it is also intricately linked to getting the maximum life out of a building and each of its operational component parts.

It is clear from the research that the hoped for 'ideal model' of building maintenance along the lines of car or aircraft servicing was not the model being pursued and is certainly not the only one for greater efficiency. Indeed, it is posited that the 'ideal' may be less efficient than the emergent JIT practice. The full promise of JIT maintenance has still to be tested, especially the consequences of repeated failure of a subcontractor to meet the contract conditions or where one incidence has major trading consequences in the retail sector. Further research will also be necessary to identify the scope for and limitations of the JIT approach to maintenance in other building types, the scale of cost savings and improvements in service, if any, and application of these findings more widely.

Summary

It has been said that 'Construction is different; it has nothing to learn from other industries'. Not only can the construction industry learn; it must do so if it is to throw off its poor image. This chapter has shown how a theory and practice from another environment can be transferred imaginatively to the otherwise staid building maintenance industry. The next chapter develops further the 'just-in-time' approach and the following chapters examine and apply further concepts from the business world.

References

American Customer Satisfaction Survey, 1995–1997. http://www.acsi.asq.org

Armstrong, M. (1994) *How to Be an Even Better Manager*, 4th edn., Kogan Page, London.

Atkinson, A.R. (1998) *The Role of Human Error in the Management of Construction Defects.* Proceedings of COBRA '98: RICS Construction and Building Research Conference, Oxford Brookes University, RICS, London, Vol. 1, pp. 1–11.

Audit Commission (1986) *Improving Council House Maintenance.* HMSO, London.

Audit Commission (2001) *Housing Inspection Service Annual Review: In Search of Excellence.* Audit Commission Publications, Wetherby, Yorks.

Bennett, J. & Jayes, S. (1995) *Trusting the Team: The Best Practice Guide to Partnering in Construction.* Centre for Strategic Studies in Construction, University of Reading.

Bottom, D., Gann, D., Groak, S. & Meikle, J. (1994) *Innovation in Japanese House-Building Industries.* Science Policy Research Unit. University of Sussex, Brighton.

Burck, C.G. (1980) *A Comeback Decade for the American Car. Fortune* **101**(11), 63.

Bushell, R.J. (1979) *Preventing the Problem – A New Look at Building Planned Preventive Maintenance.* Maintenance Information Service No. 11. Chartered Institute of Building, Ascot.

Byrne, J. (1991) Managing for quality: high priests and hucksters. *Business Week Special Issue*, **October 25**, pp. 52–57.

Cole, R.E. (1981) The Japanese lesson in quality. *Technology Review* **83**, 29–32.
Cole, R. E. (1999) *Managing Quality Fads: How American Business Learned to Play the Quality Game*. Oxford University Press, New York.
Dear, A. (1988) *Working Towards Just-In-Time*. Kogan Page, London.
Droy, J. (1986) 'JIT' for orders as well as parts. *Production Engineering* **33**, 38–39.
Duclos, L.K. & Spencer, M. (1994) *Using Information to Inform Maintenance Operations in a JIT Program: A Case Study and Analysis*. Proceedings of the 25th Annual Meeting of the Midwest Decision Sciences Institute, Cleveland, 24–26 April, pp. 210–212.
Duclos, L.K., Siba, S.M. & Lummus, R.R. (1995) JIT in services: a review of current practices and future directions for research. *International Journal of Service Industry Management* **6** (5), 36–52.
Egan, J. (1998) *Rethinking Construction: Report of the Task Force to the Secretary of State for Environment, Transport and the Regions*. DETR, London.
Finan, W. & LaMond, A. (1985) Sustaining U.S. competitiveness in microelectronics: the challenge to U.S. policy. In: *U.S. Competitiveness in the World Economy* (Scott, B. & Lodge, G., eds). Harvard Business School Press, Boston.
Gall, N. (1981) It's later than we think (interview with William J Abernathy). *Forbes* **2 February**, p. 65.
Garvin, A. (1988) *Managing Quality*. Free Press, New York.
Geertz, C. (1973) *The Interpretation of Cultures*. Basic Books, New York.
Gronroos, C. (1990) *Service Management and Marketing: Managing the Moments of Truth in Service Competition*. Lexington Books, Massachusetts.
Gronroos, C. (1994) From marketing mix to relationship marketing: towards a paradigm shift in marketing. *Management Decision* **32** (2), 4–20.
Hayashi, S. (trans. Baldwin, F.) (1988) *Culture and Management in Japan*. University of Tokyo Press, Tokyo.
Hewlett-Packard Corporation (1997) *Company Facts*. http://www.hp.com/abouthp/CorporateOverview.html.Corporate
Hofstede, G. (1980) *Culture's Consequences: International Differences in Work-related Values*. Sage Publications, Beverly Hills.
Housing Services Advisory Group (HSAG) (1980) *Organising an Effective Repairs and Maintenance Service*. Department of the Environment, London.
Latham, M. (1994) *Constructing the Team*. HMSO, London.
Lee, R. (1987) *Building Maintenance Management*, 3rd edn. Collins, London.
Marriott, O. (1967) *The Property Boom*. Hamish Hamilton, London.
McLachlin, R. (1990) *The Service Aspects of JIT Production*. Proceedings of the 1990 Decision Sciences Institute Annual Meeting, San Diego, 19–21 November, p. 1827.
Main, J. (1994) *Quality Wars*. Free Press, New York.
Mozer, C. (1984) Total quality control: a route to the Deming prize. *Quality Progress* **17**, 30–33.
Muhleman, A., Oakland, J. & Lockyer, K. (1992) *Production and Operations Management*, 6th edn. Pitman, London.
NBA Construction Consultants (1985) *Maintenance Cycles and Life Expectancies of Building Components and Materials: A Guide to Data and Sources*. NBA CC, London.
Noble, V. (1980) *The Value of Building Maintenance*. Maintenance Information Service No. 13. Chartered Institute of Building, Ascot.
Norton, B. & Smith, C. (1998) *Understanding Management Gurus in a Week*. Hodder and Stoughton, Sevenoaks.

Ohmae, K. (1981) *The Myth and Reality of the Japanese Corporation*. Chief Executive Inc., New York.

Okimoto, D., Sugano, T. & Weinstein, F. (1984) *Competitive Edge: The Semiconductor Industry in the U.S. and Japan*. Stanford University Press, Palo Alto, California.

Pajakkala, P., Matilainnen, J. & Perälä, A-L. (1993) *The Finnish Construction Branch: Markets and R & D*. Technical Research Centre of Finland, The Finnish Building Centre, Tampere.

Peters, T.J. & Waterman, R.H. (1982) *In Search of Excellence: Lessons from America's Best-Run Companies*. Harper and Row, New York, p.37.

Pettit, R. (1983) *Computer Aids to Housing Maintenance Management*. HMSO, London.

Pirsig, R.M. (1974) *Zen and the Art of Motorcycle Maintenance: An Inquiry into Values*. Morrow, New York, pp. 34–35.

RICS (1990) *Planned Building Maintenance: A Guidance Note*. RICS, London.

Riggs, J.L. (1987) *Production Systems*, 4th edn. Wiley, New York.

Ristelheuber, R. (1994) HP – America's new best-managed company. *Electronic Business Buyer* **20**, 36–41.

Scottish Local Authorities Special Housing Group (SLASH) (1979) *The Scottish Special Housing Association (SSHA) Planned Maintenance System*. SLASH, Edinburgh.

Seeley, I.H. (1976) *Building Maintenance*. Macmillan, London.

Skinner, N.P. (1983) *Interpreting Feedback Information*. Current Paper CP1/83. Building Research Establishment, Watford.

Smyth, H. J. (1985) *Property Companies and the Construction Industry in Britain*. Cambridge University Press, Cambridge.

Smyth, H.J. & Wood, B.R. (1995) *Just in Time Maintenance*. Proceedings of COBRA '95, RICS Construction and Building Research Conference, Edinburgh. RICS, London, Vol. 2, pp. 115–122.

Stalk, G. & Hout, T. (1990) *Competing Against Time*. Free Press, New York.

Trompenaars, F. & Hampden-Turner, C. (1997) *Riding the Waves of Culture: Understanding Cultural Diversity in Business*, 2nd edn. Nicholas Brearley Publishing, London.

Wheatley, M. (1992) *Understanding Just in Time in a Week*. Hodder and Stoughton, Sevenoaks.

Wilson A. (1991) *New Directions in Marketing: Business-to-Business Strategies for the 1990s*. Kogan Page, London.

Young, J. (1983) One company's quest for improved quality. *Wall Street Journal*, 25 July, 10.

4 Procurement of Building Maintenance Services

The preceding chapters have looked at some ways in which building maintenance may be provided, the focus being on the provider. This chapter explores ways in which maintenance or building care services may be obtained. The focus here is on the customer or client, the person who identifies a maintenance need and who may commission maintenance works. This person may be a building professional of some kind, perhaps a facilities manager or estates manager or building surveyor, or they may be someone with little or no experience in this area. This chapter puts the customer at the centre of the procurement process.

Context

Much maintenance work is commissioned *in extremis*. The roof is leaking and immediate attention is required, to stop it, to fix it or to minimise the effect of the water ingress. The work involved in each of those interventions may differ, perhaps significantly; the most important thing is the urgency. This puts the building user in a poor position to get good service or best value. In the domestic sector, it is uncommon for a private householder to have any ongoing relationship with a contractor. Previous experiences with a builder are likely to have been problematic. The industry is noted more for time and cost overruns and poor quality than for customer care. In these circumstances the most likely procurement route is to refer to the Yellow Pages and hope to make contact with a builder who will turn out to be satisfactory. In some parts of the world builder registration is required, for instance in the state of Victoria in Australia. A useful review of the Victoria scheme is provided by Georgiou *et al.* (2000). In the UK the title 'builder' is unprotected and listing in the *Yellow Pages* is open to anyone. All that is required to be a builder is a van and a mobile phone. Porter (1980) describes this as a 'low barrier to market entry'. Even in commercial environments, the construction industry is not always the first port of call for maintenance. A study by the author (Wood & Smyth, 1996) identified how the building maintenance market had been penetrated by the cleaning and security industries.

Further research (Wood, 1998) explored the configuration of the UK construction industry and how this impacted on housing maintenance. Rolfe & Leather (1995) identified that:

- most firms operating in the small-scale domestic repair and maintenance sectors are individuals or firms with less than six staff
- only half had a formal construction industry qualification
- few of the contractors have formal training in running a business. Some are entrepreneurial but others have been forced into self-employment by structural changes to the industry
- nearly half of those interviewed had four weeks' work or less in the pipeline (order books) at the time they were interviewed
- only 14% of contractors interviewed employed administrative or secretarial help. Most administrative work was done from home, often by wives, partners, children or other relatives, sometimes apparently on an unpaid basis.

Another review, by the Manchester Training and Enterprise Council (1997), reported the following.

- Despite its importance, the construction industry's current performance and profitability are perhaps the worst of any sector.
- The traditional culture of the industry is characterised by poor management skills and a reluctance to seek advice and help.
- Little use is made of research within the sector: there is little research into the critical market forces that influence the business.
- Increasingly clients are asking for information on the expertise of the staff allocated to their work and there is an obvious market edge for those companies that can demonstrate the competence and qualifications of their staff to do the job.
- Generally, smaller construction companies look upon networking with suspicion; invariably they see other companies as rivals.
- Few companies seek advice or help with problems or with developing the business. Use of innovative techniques or practices is rare, as is use of research and other information that might improve business performance.

A further project (Leather *et al.*, 1998) showed a range of inhibitions to home owners commissioning repair work.

- Many people buy houses without a detailed survey and most owners have no long-term maintenance plan or any special financial provision for further repair costs.
- People sometimes delay or avoid tackling work because they cannot find a builder they can trust.
- The problem of finding a trustworthy and competent builder was widespread. Owners mentioned as the main problems: major delays, poor workmanship, increased costs while on site, unacceptable behaviour, failure to clean up mess and unwillingness to return to deal with snags.
- Owners often missed complex technical problems or hidden problems.

- People rarely get professional help; they see it as expensive, hedged around with exclusions, uninformative and possibly bringing more problems to their attention than they can cope with.

A study of the construction industry in Oxfordshire (Wood, 1999) identified significant skills shortages and related problems.

- Over 40% had difficulties filling vacancies.
- Nearly half said skills needs were increasing.
- A quarter said they had a skills gap.
- Half did not use computers/IT (87% said 'don't need').
- Construction sector employees were half as likely to have received job-related education or training in the previous year compared with other industries.
- Only 30% of employers funded or arranged training.

UK governments have for some years been concerned about the poor quality of work in the domestic construction and maintenance industry. A range of reports from Simon in 1944 to Egan in 1998 was recorded in Chapter 2. The author was involved with a study in the 1980s related to quality in traditional housing (BRE, 1982). The Building Research Establishment has also produced *Defect Action Sheets* (1982–90), *Rehabilitation: A Review of Quality in Traditional Housing* (1990) and *Good Building Guides* (1990–). The government initiative 'Combating Cowboy Builders' (DETR, 1998) was referred to in Chapter 1. A paper by the author in 1998 quoted from the related consultation about rogue traders.

> 'Latest figures from the Office of Fair Trading (OFT) in the year to 30 September 1996 indicate over 93,000 complaints over home maintenance ... up 4% on the previous year, a trend very much in line with the previous ten years which has seen an average rise of 5% a year. By way of comparison, it is also considerably larger than the number of complaints received about second-hand cars (approx. 84,000).'

While pilot schemes to improve the situation have been introduced in Birmingham and Somerset, evidence of improved practice remains hard to find. The DETR Annual Report for 2000 showed a decrease in quality with increasing instances of defects. It seems that the 'cowboy' may be an enduring feature of the building maintenance market.

Contracts and relationships

> 'The contractual arrangement adopted determines not only the relationship of the various design, construction and advisory organisations with the client but also their relationships with each other.' (Nahapiet & Nahapiet, 1985)

Clear contractual arrangements and associated documentation, with the obligations of all parties spelled out, should enable risks to be clearly assessed and costed. Higher levels of uncertainty generally generate higher levels of pricing. This has given rise to the development of standard forms of contract, particularly the Joint Contracts Tribunal (JCT) series, designed to cater for as many problems as could be anticipated, providing mechanisms for resolution, and amended periodically as new 'case law' is generated.

Traditionally for new-build projects, the lump sum contract awarded by single-stage selective tender has been the common procurement route. A shortlist of contractors capable of carrying out the project to the quality and in the timescale required is invited to tender - typically six (the actual number perhaps being guided by advice of the Joint Contracts Tribunal - JCT, 1994), of which all but one will be 'losers'. Of course, it is possible that the 'winner' will also become a loser if that tenderer has seriously underestimated the cost of the work or was desperate for more work at that time. In this case the client will also be a loser if the contractor cuts corners or fails.

Alternative forms of procurement have been tried, such as:

- negotiated contracts
- serial contracts
- management contracting
- construction management
- design-build
- design-build-finance-operate (DBFO)
- build-own-operate-transfer (BOOT)
- public-private partnership (PPP).

However, there is still great reliance on the traditional relationship. Institutions, particularly with their standing orders, are often constrained to use selective tendering in pursuit of openness and value for money (often misinterpreted as cheapness).

By contrast, with maintenance work, because of its inherent uncertainties in terms of scope and scale of the work, the unpredictability of when emergency work may arise and complications of access, often involving disruption to occupants and their operations, contractual arrangements have often been looser. Maintenance 'projects' may be as small and short-lived as the repairing of a leak or replacement of glass, with a 'contractor' identified by a householder by reference to the *Yellow Pages* and with the 'contract' being little more than acceptance of a verbal estimate. In essence, these 'call-outs' are dealt with as 'daywork', generally favourable to the contractor, with the client having little control over cost, quality or timescale. Perhaps surprisingly, systems not very dissimilar from this have also existed in-house within companies or corporations as direct labour or direct service organisations (DSOs). Sometimes client organisations have developed relationships with contractors to provide such 'call-out' services as and when required. These arrangements have developed because they made sense. On the one hand, why go 'outside' to contractors, contributing to their profits when these could be retained 'inside'? On the other hand, why retain staff on the books, with associated employment and accommodation costs, if the service can be obtained satisfactorily outside? In either case, why incur the costs and delay of selective tendering for every job? The author's Montreal paper (Wood, 1997) averred that: 'Such arrangements thrive when there is trust between the parties; they die when trust is abused or accountability is questioned'.

In the maintenance sector, where formal contracts exist, Measured Term Contracts using either standard (e.g. National or PSA Schedules of Rates) or bespoke pricing and payment systems have become a common arrangement, particularly for institutional clients. With the contract enduring over a number of years, this allows for the development and nurturing of mature, trusting relationships between the parties.

Service and 'service culture'

The construction industry is not noted for its attention to customer needs. Arguably, if it were, it would give more attention to completion of projects on time, within budget and without defects. For a long time there was no client representation on the Joint Contracts Tribunal, the organisation responsible for the drawing up and revision of the standard form of building contract used in the UK and much of the former British Empire. The architect was ascribed a 'quasi-arbitrarial' role in which he or she would determine issues related to defects and delays, extra costs and extensions of time. Once construction started, no reference was required to be made to the client, whose job was only to pay up and put up, and to accept the building as and when built. A client's presence on site would be seen as an unwelcome nuisance and a contractor's interest in keeping the client happy would be primarily motivated by hoping to be asked to tender for a subsequent project. A cynic would suggest that little has changed.

Who procures?

Notionally it is the client that procures the building or service but that is to oversimplify the matter. The client may believe that a new building is required, for instance to cope with an expanding business. Equally this perceived need could be met, in part or in whole, by the extension or adaptation of an existing building which may or may not be in his or her possession. Perhaps business operations could be reconfigured such that no new building is required; indeed, it may be that less space will be required in the future.

What information and skills will a 'typical' client have to enable such crucial decisions to be taken? Some matters are very much building or property related and may be within the normal capabilities of an architect or surveyor to advise upon, in which case the issue for the client is largely about how to select the most appropriate. Concerns will normally relate to location and size of the firm and its expertise or 'track record' with the kind of development envisaged. However, these professionals may not be particularly well placed to advise on business processes and how these may inform building needs. Good financial advice is also required. These people must work effectively as a team to identify and evaluate a range of alternative options and to advise their client on the best procurement route. There is much literature on selection of procurement routes (e.g. Nahapiet & Nahapiet, 1985; Masterman, 1992; Morledge & Sharif, 1996) and it is not intended to discuss that further here. Suffice it to say that the professionals advise and the client acts upon that advice.

Who is the client?

This can be complex. In one part of the spectrum are major clients with large programmes of ongoing work, such as the large retailers (Marks and Spencer, McDonalds, Sainsbury, Tesco) who may have several hundred stores and millions of square feet to build or refurbish every year. Then there are government departments such as the UK Ministry of Defence and quasi or previously governmental bodies such as the BBC, BAA (formerly the British Airports Authority), utility companies (gas, water, electricity, telecoms) and the NHS. Many of these bodies have been at the forefront of innovation and good practice in procurement. The BBC was named 'Construction Client of the Year, 2002' (Baillieu & Black, 2002); Sir John Egan was Chief Executive of BAA plc. At the other end of the spectrum are individual householders and in between are a variety of small landlords, corporate bodies such as schools, colleges and care homes, and companies of various sizes; many of these will have no property-related or technical expertise.

Commissioning of maintenance work

Who will procure what, when, how, how often and why?

Many of the larger organisations will have a maintenance manager, though their job title may be property manager or estates manager, buildings officer, facilities manager or similar. Many will have teams, units or departments involved in commissioning and maybe carrying out maintenance work. Such work may be carried out to programmes and priorities determined by themselves; there may be policies and procedures to work to, which have been approved by a board or committee or directorate. Some may have standing arrangements with consultants such as building surveyors and others with a general building or maintenance contractor. Some will have no arrangements in hand to deal with maintenance. Whether there is a dedicated staff will depend very much on size, distribution and quality of estate and company philosophy and what that means in terms of quantity and quality of maintenance required.

A headquarters building, much visited by clients and potential clients, is likely to suggest a high level of service, at least in the entrance and public areas, meeting rooms and visitors' toilets. 'Back office' areas may be given a lower level of attention and production and storage areas lower still. This is not to justify that order of priorities; on the contrary, there is much to be said for treating staff genuinely as 'our most valued resource', maintenance of the building being seen as an investment in people's working conditions. It is possible to assess the level or quality of care by comparing maintenance expenditure with building asset/replacement value; figures of 1.5–2% of asset value have been recommended. It is suggested here, however, that a more people-centred paradigm would promote measurement of quality by amount of investment per person, perhaps even by proportion of user expectation actually met.

Priorities: time/cost/quality

Different organisations, and different individuals within an organisation, will have differing expectations, perhaps widely varying. Some will place the 'standard' of maintenance high on their list of priorities; others will prefer to see time and money invested in other areas. This can create difficulties for maintenance providers trying to meet differing and perhaps poorly expressed demands. Certainty must be sought if it is not provided. Documentation can help, especially if it conforms to a consistent or standard pattern, although there is also much to be said for development of a trusting, long-term relationship in which the contractor provides unquestioningly professional service. Such trust, common in many Japanese companies, was considered in Chapter 3 and further informs discussion in subsequent chapters.

It is common to regard priorities of time, cost and quality as being in conflict and that choices have to be made between them. Often this is represented diagrammatically as a triangle, with T, C and Q either at the points or alternatively as the sides of the triangle, sometimes called the eternal, or infernal, triangle. A project is then defined by its positional proximity to the three dimensions. Here are three possibilities.

(1) It is vital the project is finished on time (better still, as soon as possible), whatever the cost; quality is unimportant as long as the building stands up and keeps the water out.
(2) Completion within budget is essential and industry-standard specifications of quality must be met. If the project is ready late, so be it.
(3) The project must be completed on time and within budget; lower specifications will be adopted for certain areas, if need be.

Assigning priorities

There is a wide variety of positions within and between these priorities. What the model does is to assist the client to determine priorities and to express these accordingly to consultants and contractors. The model does, however, have limitations. It assumes 'trade-offs' between criteria on the basis that not all criteria can be met simultaneously; that something must be sacrificed. It is difficult therefore to place within the triangle a project that seeks delivery of a high-quality building at low cost within a short time. Perhaps some thinking outside the triangle is required.

It is important for the client to 'get a grip' on the relevant priorities, otherwise they will have to put up with the service provided. In the absence of clear instructions, it is sensible for a contractor to seek clarification before commencement, though that is not easy when work is ordered *in extremis*, as is common with building maintenance. This is where a relationship of mutual understanding and trust is valuable. Then the contractor knows instinctively, if not explicitly, the kind of job required: whether it is more important to 'get it fixed ASAP, no matter what it looks like' or 'if it takes six months to find a matching piece of marble, then that's how long it will take'. Professional contractors - the only kind we really want to deal with - have a lot to offer clients, especially the smaller and less experienced client. For a continuing and deepening relationship they will want to check out with the client the kind of response preferred. Which would you prefer?

- I did what you asked
- I did what was best
- I did what we agreed

Developing the client–contractor relationship

Recent years have seen the creation and growth of the facilities management function providing or procuring services relating to the care of buildings beyond their construction. Traditionally, construction companies' interest in building projects was restricted to the period from tender to completion. As indicated in the previous chapter, it is very rare for building contractors to develop long-term relationships with clients. Clients have become increasingly aware of the running costs of buildings and the scope for reducing these by appropriate design and construction. A thoughtful design could both reduce the amount of maintenance and also make it easier to carry out the work. Contractors are well placed to offer 'cradle-to-grave' services in relation to buildings but are often not capitalising on this potential and indeed, have been yielding existing markets to operators from other fields.

The development of design & build and similar procurement methods has brought the involvement of contractors forward into the design stage. Maintenance of buildings has generally been carried out by multiskilled 'handymen', by direct labour forces employed by public clients and by small builders, especially where small-scale adaptations have accompanied repair works, and generally on an *ad hoc* basis, often with relatively light contractual arrangements. Maintenance could be improved by greater attention at the design stage.

The Private Finance Initiative (PFI) is giving impetus to build & operate or design/build/operate contracts whereby contractors and/or their financiers will gain or lose projects and profits according to how well they can forecast and contain running costs. This is likely to give rise to increasingly sophisticated maintenance planning and procedures and clients will need professional consultant and contracting services in these areas.

Historical development

The 'scientific' approach suggested by PPM and LCC has often failed in practice because sufficient funds were not available, because maintenance is not a high-profile activity and because maintenance managers also have been lowly valued.

In-house maintenance managers and staff were not rewarded if they endeavoured to pursue the PPM ideal because the benefits could take some years to show through, if indeed they were noticed, and they are largely unquantifiable. There has been an absence of criteria for evaluation of in-house staff. More substantial maintenance and refurbishment work has typically been put out to construction companies. For these contractors the source of work was similar to new-build work, although anecdotally it could be argued that there was a higher incidence of repeat business.

However, the growth of facilities management (the first International Congress of the International Facilities Management Association (IFMA) took place in 1989) and the increase in interest of the 'traditional' building professions represented by, for instance, the RICS, CIBSE and CIOB has elevated debate about maintenance management. As Michel put it in his foreword to the *CIOB Handbook of Facilities Management* (Spedding, 1994), '... facilities management has begun to be recognised as a key element of an organisation's strategic plan, with responsibility for its effective pursuance firmly located within the boardroom'.

These changes in management outlook have taken place against a background of opportunities created by information technology and communications networks. Lee (1987) wrote that 'few technologies have developed at the speed of computer design or are as likely to have such a fundamental effect on our everyday lives'. He reminded us that 'the first electronic computers were huge contraptions requiring special air-conditioned rooms ... then, in the sixties ... computers the size of one or two filing cabinets ... now, the desktop microcomputer and even smaller portable models'.

Since then speeds and capacities of computers have increased even more, to the point where information on whole stocks of buildings can be accessed across extensive networks, updated by inputs from handheld and other remote devices. The incorporation of systems including sensing and control devices and connections into properties has realised the possibility of the 'intelligent building'. The technologies available, especially if built into the building, and management systems which may be directed by building owners now play large parts in the procurement of building maintenance services. 'Intelligent building maintenance' is covered further in Chapter 7. The market has become very much customer driven and building contractors and consultants need to respond.

The service context: building relationships

It is not intended to suggest that 'cosy' arrangements between clients or their agents and contractors chosen on merit rather than lowest price are corrupt or that they represent poor value. However, a developing 'professionalisation', together with the changing managerial context of organisations working through concepts such as 'return to core business', 're-engineering', 'downsizing' and 'outsourcing', has brought the cost of property, including maintenance, into the spotlight.

Term contracts have brought benefits of scale to clients and contractors, though often at the expense of voluminous paperwork in their preparation and execution. Speakers at a meeting of the Construction Industry Computer Users Group (CICUG, 1996) bemoaned the tendency of clients, especially local authorities and large commercial companies, to devise their own systems with bespoke Schedules of Rates using individual definitions, conventions and specifications. This has made it difficult for contractors to price, even when they have built up good records of costs to inform unit pricing. While term contracts offer the benefit of continuity of work for contractors, and for clients, the larger contracts of themselves mean fewer opportunities to bid for work, especially for smaller firms, and a bigger 'downside' to losing contracts with existing clients. This points up the increased importance of creating and sustaining good client–contractor relationships.

Building contractors generally have low levels of repeat business of around 20%, compared with levels of 60–80% in other industries, and tend to practise vigorous sales strategies focusing on price cutting to achieve competitive advantage (see Porter, 1980) rather than relationship marketing (Gronroos, 1990, 1994). Loyalty and trust do not feature prominently in construction (Smyth, 1997; Thompson, 1998) and partnering is a recent phenomenon (Latham, 1994; Bennett & Jayes, 1995) which is being implemented with variable degrees of coherence.

Contractors are also seeing building maintenance work being awarded to companies based in or coming from other sectors, such as security, cleaning or catering, with clear customer orientations and used to meeting expectations of service quality and reliability not always associated with 'builders'. It may be that construction companies able to differentiate themselves from the rest will be able to generate market share and repeat business based on sound business relationships.

Developing and defining just-in-time maintenance

The development of the intelligent building, including highly serviced offices and supermarkets for instance, brought with it the potential of just-in-time maintenance (Smyth & Wood, 1995), as identified in Chapter 3.

JIT maintenance may be characterised by features such as:

- inclusive (all-in) service
- customer focus
- service level agreements
- contractor input
- modularised equipment.

In essence, JIT maintenance is defined as: 'getting the maximum life from each building component and piece of equipment, leaving repair or replacement until the component is broken or fails to function, yet taking action prior to it having a serious effect upon the performance of the organisation'. In the retail case this is prior to it affecting customers, stock and hence profit in a detrimental way. It is achieved through a rapid and flexible response from the subcontractor.

A maintenance system, whether JIT or not, is capable of producing mountains of information on performance, reliability and response, which can inform decisions on specification and purchasing of equipment. Feedback data on cost in use provides input to future lifecycle costing exercises, as long as it is not based on equipment which may already be obsolescent and where standards for the future may differ.

Similarly, such data could inform appropriate periods for planned preventive intervention. As Wild (1995) has identified in relation to production and operations management:

> '... we want to perform the minimum amount of preventive maintenance, since maintenance even of this type will be expensive in terms of labour and material costs, and also the possible costs of disruption. Ideally, therefore, we would like to perform our preventive maintenance of equipment just before it would otherwise have broken down. Such a policy is possible only if either, because of the nature of the equipment, we receive some advance warning of impending failure, or if failure of equipment is perfectly predictable.'

Wild goes on to contend that 'rarely is warning of impending failure of value in practice, since either the warning is insufficiently in advance of failure, or the warning is itself associated with some loss of efficiency or capacity in the equipment'.

Performing to standard

The research on supermarkets showed that prediction is becoming increasingly good, or good enough, to allow inspection to be reduced or eliminated and service intervals tuned in the light of specific equipment experience in service. It also showed that warning can be received sufficiently in advance of failure for an appropriate response to be made. This has provided the maintenance contractor with the remit and resources to meet the whole of a store's maintenance needs, each job within a specified response period from time of 'order' and for a known sum. Contracts have moved from specifying service intervals to specifying performance standards. Furthermore, contractors are able, from their experience on such contracts, to negotiate the setting of appropriate and achievable standards of performance and to influence future equipment provision. Install-maintain, or design-install-maintain, may be expected to become more common.

It was found also that where contractors were unable to provide a full service from their own resources, it is common to subcontract, just as in building work, and that for some specialist elements of service, such as maintenance of weighing equipment, nomination is the norm.

Setting JIT performance goals

Increasing use of modularised equipment, such that parts may be readily replaced, maintaining or restoring service as quickly as possible, is helping to eliminate the downtime and reduce disruption, which is unacceptable in the supermarket sector. These are stringent demands with seven-day trading and long opening hours, a situation increasingly common across a range of sectors in the global economy.

A problem identified with the system arises from its very efficiency. There are situations where the speed of response called for in the contract documents may be inappropriate, for instance around a Bank Holiday, or where there is 'manual intervention,' for instance from a departmental manager or regional engineer and conflicting or repeat requests are received at the service desk. Someone needs to negotiate priorities. It is useful to have a limited range of response times from the lower limit of two hours for outsourced JIT maintenance to a maximum of seven days. Thus, control is centralised while the response is highly decentralised.

Implications for building contractors and consultants

Changing maintenance needs and procurement systems bring both threats and opportunities to the traditional suppliers of services. The study of practices in supermarkets showed that maintenance services were being provided by organisations from a range of backgrounds:

- building contractors
- services engineering contractors
- equipment manufacturers
- security and cleaning companies.

Analogies can be drawn with the growth of the facilities management industry, with a common feature being the 'one-stop shop', the single point of contact, with a growth in larger, longer term contracts in which small firms or individuals may provide services 'as and when required' on subcontract. There are thus implications in terms of seeking, winning and sustaining contacts and contracts over a long period of time. This prospect of continuous workflow brings with it the possibility of creating and the means of providing a highly disciplined workforce, in some ways similar to the opportunities offered by prefabrication. In manufacturing industries these would be termed long production runs and would be highly sought and valued.

There are implications in terms of marketing and investment. Contractors particularly need to be able to present potential clients with a certainty of delivery so that the client may be confident of service in turn to their tenants. This trust takes time to build, though it can be quickly eroded by underperformance or overselling. Clients will also want to be assured that the contractor is committed to them and to the sector – this will need investment particularly in the service infrastructure.

Built-in maintenance

Building contractors are already facing competition in the domestic market from trusted organisations like the Automobile Association and Green Flag Insurance using their communications infrastructure with national and international coverage linked to certified, trained personnel and guaranteed delivery. This is discussed further in subsequent chapters.

It may be that contractors will be able to secure maintenance work for themselves by 'building it in' at the design or construction stage, not by the creation of defects but by suitable specification and contractual agreements. The Private Finance Initiative is giving impetus to build & operate or design/build/operate contracts whereby contractors and/or their financiers will gain or lose projects and profits according to how well they can forecast and contain running costs. 'A skilful design can reduce the amount of maintenance and also make it easier to carry out the work' (How Son & Yuen, 1993). Contractors will also have opportunities to specify and install equipment and systems that they will find easy and economical to maintain, from manufacturers or suppliers with whom they may develop long-term relationships; installations may require maintenance by 'approved firms'. By contrast, of course, more clients, and consultants, may require contractors to install systems specified by the security, cleaning and catering companies. Consultants may also find roles in setting up and monitoring service contracts and at the design stage in specifying appropriate 'smartware' systems, performance standards and response times.

Implications for building owners and users

Building owners will need to review their briefing procedures for new buildings, their management and maintenance. It is now both practical and desirable to adopt a holistic approach to the building. Design, specification and construction can reasonably be expected to take full account of the maintenance requirement and subsequent upgrading needs. There may be real benefits, including cost benefits, to be gained by engaging a contractor who will take on a 'lifetime' role and relationship, in which case the building owner will need to weigh his relative power, including purchasing power, against potential contractors or building suppliers. The contractor may be offering a building equipped by his or her preferred suppliers with systems not fully compatible with systems already in use or planned in other buildings in the client's portfolio. Perhaps the building owner will be tied into systems and suppliers that are or become expensive to maintain or to upgrade.

Maybe purchasing decisions for the intelligent buildings of the future will be similar to those relating to today's computing, where hardware, software training and support need to be considered very much together. This implies a high level of technical, project management and financial skills in the building, property or facilities manager or surveyor of tomorrow (Stacey & Wood, 1996).

These surveyors will also need to maintain good working knowledge of contracts and contractors, their strengths and weaknesses and decision-making abilities as to which to use, when and for how long. At present skills are required in estimating the scale of repair work and attributing priorities and likely costs. Monitoring devices together with decision support systems (formerly known as expert systems) will increasingly 'automate' these processes. This will move the focus of surveyors' skills to budgeting, so that sufficient funds are available for the JIT maintenance works, and to the creation of a contractual milieu that is conducive to a high quality of service.

For users there are implications in terms of control. Studies have shown that the imposition of remote control systems can be poorly received by building occupants. For instance, Haves (1992), in the context of building energy management systems (BEMS), has shown a preference for occupants to be in control of their environment, rather than have their heating, lighting, occupation rates and utilisation controlled or monitored remotely, whether by sensors or 'the controller'! There are possible implications for individual loss of privacy – perhaps 'Big Brother' is watching.

Conclusion

Growth in IT systems for managing building maintenance has been paralleled by the development of devices and systems for managing the buildings themselves. Computing power is such that it is possible to forecast when building services equipment or fabric are likely to fail and to facilitate, indeed to order, appropriate JIT maintenance action.

A significant requirement for the success of JIT maintenance is the provision and sustaining of an infrastructure of suitable systems and personnel. This provides the opportunity for building contractors to develop long-term relationships with clients based on procurement methods and service qualities with which they may be relatively unfamiliar. Providers more familiar with the terminology and practice of service delivery are already operating in this field, having developed their skills elsewhere.

Building contractors presently providing maintenance services need to consider their market and take both more proactive and more responsive action to identify and meet the needs of their present, past and future clients. All participants, clients, consultants and contractors, have the opportunity to propose and ultimately impose the procurement methods most appropriate to them.

Summary

This chapter has looked at a number of ways in which building maintenance services may be obtained. It has considered how, for instance, the growth of IT capabilities has opened up new possibilities, enabling both providers and procurers of maintenance service to re-evaluate 'traditional' procedures. The possibility of just-in-time maintenance has been identified. The next chapter goes further, suggesting a deeper, more fundamental 'root-and-branch' examination, re-engineering processes from first principles.

References

Baillieu, A. & Black, S. (2002) The top 50 clients. *RIBA Journal* **109** (9), 24–30.

Bennett, J. & Jayes, S. (1995) *Trusting the Team: The Best Practice Guide to Partnering in Construction.* Centre for Strategic Studies in Construction, University of Reading.

Building Research Establishment (1982) *Quality in Traditional Housing*. HMSO, London.

Building Research Establishment (1982–90) *Defect Action Sheets.* HMSO, London. (Published 1991 as *Housing Defects Reference Manual.* E. & F. N. Spon, London.)

Building Research Establishment (1990) *Rehabilitation: A Review of Quality in Traditional Housing* (Report BR 166). BRE, Watford.

Building Research Establishment (1990–) *Good Building Guides*. BRE, Watford.

Construction Industry Computer Users Group (1996) *Property Maintenance and Services.* CICUG, London.

Department of the Environment, Transport and the Regions (1998) *Combating Cowboy Builders: A Consultation Paper.* DETR, London.

Georgiou, J., Love, P.E.D. & Smith, J. (2000) A review of builder registration in the state of Victoria, Australia. *Structural Survey* **18** (1), 38–45.

Gronroos, C. (1990) *Service Management and Marketing: Managing the Moments of Truth in Service Competition.* Lexington Books, Massachusetts.

Gronroos, C. (1994) From marketing mix to relationship marketing: towards a paradigm shift in marketing. *Management Decision* **32** (2), 4–20.

Haves, P. (1992) Environmental control on energy efficient buildings. In: *Energy Efficient Buildings* (Roaf, S. & Hancock, M., eds). Blackwell Science, Oxford.

How Son, L. & Yuen, G.C.S. (1993) *Building Maintenance Technology.* Macmillan, London.

Joint Contracts Tribunal (1994) *Code of Procedure for Single Stage Tendering.* JCT, London.

Latham, M. (1994) *Constructing the Team.* HMSO, London.

Leather, P., Littlewood, M. & Munro, M. (1998) *Make Do and Mend: Explaining Home-Owners' Approaches to Repair and Maintenance.* Policy Press, Bristol.

Lee, R. (1987) *Building Maintenance Management,* 3rd edn. Collins, London.

Manchester Training and Enterprise Council (1997) *Building Success: A Guide to Best Practice for Small Businesses in the Construction Sector.* Manchester TEC, Manchester.

Masterman, J.W.E. (1992) *An Introduction to Building Procurement Systems.* E.& F.N. Spon, London.

Morledge, R. & Sharif, A. (1996) *The Procurement Guide.* RICS Books, London.

Nahapiet, H. & Nahapiet, J. (1985) A comparison of contractual arrangements for building projects. *Construction Management and Economics* **3**, 217–231.

Porter, M.E. (1980) *Competitive Strategy: Techniques for Analysing Industries and Competitors.* Free Press, New York.

Rolfe, S. & Leather, P. (1995) *Quality Repairs: Improving the Efficiency of the Housing Repair and Maintenance Industry.* Policy Press in association with the Joseph Rowntree Foundation, Bristol.

Smyth, H.J. (1997) *Partnering and the Problems of Low Client Loyalty Incentives.* Proceedings of the 2nd National Construction Marketing Conference, Oxford, July, pp.10–17.

Smyth, H.J. & Wood, B.R. (1995) *Just in Time Maintenance.* Proceedings of COBRA '95: RICS Construction and Building Research Conference, Edinburgh. RICS, London, Vol. 2, pp. 115–122.

Spedding, A. (ed.) (1994) *CIOB Handbook of Facilities Management.* Longman, Harlow.

Stacey, C.N. & Wood, B.R. (1996) *Surveying Needs.* Proceedings of COBRA '96: RICS Construction and Building Research Conference, Bristol. http://www.uwe.ac.uk/conferences/COBRA96/cobra62.htm.

Thompson, N.J. (1998) *Can Clients Trust Contractors? Conditional, Attitudinal and Normative Influences on Clients' Behaviour.* Proceedings of the 3rd National Construction Marketing Conference, Oxford, 9 July, pp. 7–15.

Wild, R. (1995) *Essentials of Production and Operations Management,* 4th edn. Cassell, London.

Wood, B.R. (1997) *Building Maintenance Service Procurement.* Proceedings of the Conseil Internationale du Batiment et du Recherche (CIB), Working Commission W92, University of Montreal, June, pp. 801–811.

Wood, B.R. (1998) *Maintenance Service Development.* Proceedings COBRA '98: RICS Construction and Building Research Conference, Oxford, July. RICS, London, Vol. 2, pp. 169–177.

Wood, B.R. (1999) *Construction Skills Training in Oxfordshire.* Centre for Construction Management, Oxford.

Wood, B.R. & Smyth, H.J. (1996) *Construction Market Entry and Development: The Case of Just in Time Maintenance.* Proceedings of the 1st National Construction Marketing Conference, Oxford, July, pp. 17–23.

5 Re-engineering the Process

Previous chapters have considered the development of a range of approaches to the procurement and provision of building maintenance services from the perspectives of building owners and their professional advisers, on the one hand, and contractors on the other. This chapter suggests the application of the 're-engineering' philosophy of Hammer & Champy to deconstruct traditional, evolutionally developed, 'tried-and-tested' approaches, to question received wisdom and to fundamentally rethink from 'square one'. Subsequent chapters develop further some of the outcomes of that review. The aim is to promote radical review of current practices, by questioning underlying assumptions and thinking afresh.

Hammer & Champy

Chapter 3 presented an overview of the development of 'management' through the 20th century. Industry had passed through a period of growth by acquisitions and mergers, creating huge diversified conglomerates such as Distillers, the Hanson Group and Trafalgar House, and was about to move into reverse with a return to core business, demergers and downsizing. In 1990, Michael Hammer, former computer science professor at the Massachusetts Institute of Technology (MIT), published 'Re-engineering work: don't automate, obliterate'.

Hammer's simple thesis was that in order to make substantial gains in efficiency it was necessary to go back to first principles; it was insufficient to merely automate existing inefficient processes. Through radical 'root-and-branch' reviews of what was really required, it may be possible to reduce or eliminate whole procedures. In 1993, Hammer was joined by James Champy in co-authoring *Re-engineering the Corporation*, which extended the principle to reviewing the entire organisation.

Critiquing 're-engineering'

In appraising *Re-engineering the Corporation*, Norton & Smith (1998) identified a definition of 're-engineering' as 'the fundamental rethinking and radical redesign of business processes to achieve dramatic improvements in critical contemporary measures of performance such as cost, quality, service and speed'. They go on to suggest that re-engineering 'fails or at best produces only marginal results in the majority of organisations in which it is implemented', positing that 'programmes are not sufficiently radical, only tinkering'. It is also suggested that it may be 'because the human side is not managed sensitively and re-engineering is used as a cover for making redundancies'.

The rhetoric of re-engineering tends to be hard and aggressive. There will be suggestions of cutting, eliminating, halving, slashing, stripping out waste. More sensitively, but intending to amount to much the same, paring, reducing significantly, saving or trimming may be more acceptable, 'softer' expressions. As Armstrong (1994) sees it, business process re-engineering (BPR) 'tends to be an all-or-nothing proposition ... not worth doing unless there is a pressing need to rethink what the organisation is doing overall or in a major area'.

Armstrong cites a number of examples of successful implementation of BPR.

- A Glasgow engineering firm, by focusing on the core competences required to meet new business demands, introduced multidisciplinary teams, reducing management layers from nine to four.
- Ford re-engineered its accounts process, introducing invoiceless processing which achieved a 75% reduction in headcount.
- Mutual Benefit Life reduced its turnaround of customer applications from 5–25 days to 2–5 days, eliminating 100 jobs.
- IBM Credit cut its time for preparing quotes from seven days to one.
- Bell Atlantic cut its delivery times from 15 days to one.

Because BPR is intended to produce radical and extensive changes, these can be difficult to implement in a gentle, phased way. It is not only 'all or nothing' but also 'all at once'. This can give rise to severe culture shock or what Toffler (1970) called 'future shock' and serious resistance to change.

Aggressive language

Re-engineering is characterised by words such as:

Fundamental	Revolution
Rethinking	Dramatic improvements
Radical	Critical
Redesign	Organisational change

Elsewhere in this chapter you will find 'drive down', 'dramatically reduce', 'slash', 'win', 'substantial increase', 'cut', 'eliminate', 'halve' and 'strip out'. Such words, and the sentiments they encapsulate, represent a 'macho' style of management that may appeal to those who seek the 'smack of firm government'. However, it can involve a scale and rapidity of change that it is difficult for others, downstream of such decisions, to respond to with alacrity or enthusiasm, especially if they were not involved in the decision-making process. By contrast with the 'if it ain't broke, don't fix it' school of thought, it has been suggested that re-engineering demands a more interventionist approach that may be described as 'if it ain't broke, break it'.

Middle way forward

Perhaps there is scope for a 'middle way' that brings together a radical review with a more gentle, gradual approach to the introduction and implementation of change. Current practice would be picked apart, going right back to the basics of why things are done at all and not just how they are done or why they are done in the way that they are. Coupled with perhaps radical brainstorming, thinking the impossible, this could give rise to a number of possible ways forward. These in turn can be subjected to all manner of assessment and evaluation. It may be that no new way is deemed to offer sufficient benefit to be worth the scale of investment or risk involved in making the change. Change is a chancy business.

People are on the whole resistant to change; it is challenging and uncertain. It will take resources to plan and introduce the change and it may not work out. Arguably, the bigger the change, the larger the costs, the higher the risk and cost of failure. Britain has seen many examples over recent years of large computer and document handling projects that failed to deliver, for instance in the health service, housing benefits, criminal records, passports and the Stock Exchange. The record for construction projects is not good either, in terms of time, cost and quality.

Of course, it is always possible to secure improvement in outcomes by redefining them. For instance, a project would be more likely to be completed 'on time' if an extra three months is added to the duration otherwise thought reasonable; and 'within budget' the larger the contingency sum added. The punctuality of Britain's trains improved overnight by adjusting from five minutes to ten minutes the amount of lateness within which a train would be assessed as being 'on time'. There is an argument, however, that the more time there is available, the later the job is started or the longer it takes.

Implementation

Timescale, training and commitment are critical considerations. A project scheduled for completion over a short period runs the risk of having been underestimated in both time and complexity, with the resulting stress of trying to 'achieve the impossible'; there is no time in which to respond to and recover from problems. In a short project, however, there is likely to be a team spirit that will help to gain and maintain a momentum, a will to overcome problems and win through. By comparison, a long project gives the opportunity to see how things are going and to change plans, deploying and redeploying resources accordingly. There is, however, a possibility that it will be more difficult to build up a head of steam and to generate drive and impetus.

Training is also important for both the smooth execution of the changeover project and the continuing operation of the new processes. The British construction industry, although well served by the Construction Industry Training Board (CITB), one of the last two government-supported training boards, is not good at consistent, ongoing training. The CITB focus is very much on securing and sustaining a flow of new apprentices into the construction trades. Despite good, authoritative forecasts at national and regional levels, it is for individual construction companies to decide on their own corporate and individual skill deficiencies and related training needs. There is comparatively little resource allocated to ongoing development and upskilling. A study by the South East of England Development Agency (SEEDA, 2002) identified construction as having the lowest proportion of people receiving recent training.

Commitment at both organisational and individual level is imperative. It may be argued that a proposal that comes 'from the bottom up' is more likely to succeed, through the commitment of those responsible for the operation of the new processes, than one imposed from 'on high'. Involvement in the re-engineering process, the analysis and design of new processes, by all who will be involved in making the new way a success is important. Consultation exercises will be of limited help if they are perceived as being hollow pretences at true participation. There is much to be said for a small pilot study, introducing the new ideas and ways of working in a part of the organisation. This should be monitored and evaluated and modified as necessary before 'rolling out' more widely.

Incrementalism

The principal advantage of re-engineering is in the radical reviewing of what is done and why. This total review need not imply total revision, nor a 'big bang', comprehensive, all-in-one-go imposition of new systems. The merit of a pilot study has been asserted; there is similar value in an incremental, 'step-by-step' approach to introduction of change. A programme will have much to commend it if it:

- balances demands for new learning over time
- recognises and addresses people's fears
- has a 'dry run'
- includes regular monitoring and review
- is seen to deliver good results, and preferably soon
- allows for individuals to be valued and supported
- stretches without being overambitious
- can be dropped or rescheduled, returning to known ways if necessary.

Such a programme promotes institutional and individual development and learning. Its nature is to facilitate and support the Japanese concept of kaizen – continuous improvement – through a parallel process of continuous learning. The approach is congruent with such concepts as 'experiential learning' (Kolb,

1984), the 'reflective practitioner' (Schön, 1987) and the 'learning organisation' (Senge, 1990).

Amongst the gains to be achieved by such an approach are the following.

- Radical review prompts a fresh look with no 'sacred cows', although this can produce a feeling of threat to the original proponents and operators of current systems and consequential defensiveness.
- A recognition of the foregoing feelings and therefore an attention to emotional matters rather than a marginalising or denigration of such responses.
- Commitment of all currently involved in suggesting possible ways forward and 'signing up' to proposed improvements.
- Greater likelihood that expected improvements will be achieved and full involvement of participants in trying to understand what went wrong if improvement was not achieved or was less or later than expected.
- Through continuous or periodic review, increasing realism about what can be expected, possibly resulting in more modest but realistic objectives for subsequent programmes. Also recognition of the possibility of higher aims being achievable and the resources required to so achieve.
- Identification and articulation of priorities.
- Organisational improvement and individual development, enhancing opportunities for future development.

Some of the above allow for measurement of degrees of improvement, although it is important not to downgrade or downplay qualitative matters in favour of the quantitative. Thus, while it is possible and important to measure response times such as time to pick up and answer a telephone call or to get the relevant tradesperson to the breakdown, it is also important to measure the degree of satisfaction of the service user. This can be done using an appropriate and consistent scale. Improvements, or declines, in service quality over time can be plotted and prioritised for action within future re-engineered services.

Such aspects of service that lend themselves to measurement could be automated. Sensors and other devices that may be construed as constituent parts of a so-called 'intelligent building' could report on performance as part of their function. If this were linked to measurement of a change in performance consequent upon a recorded action, this would offer the possibility of something akin to 'learning' through the building up of a database which could help inform and activate future actions – the 'learning building' (McGregor, 1994).

Background to re-engineering building maintenance services

This chapter will reflect on recent developments in the ways in which building maintenance services are being configured, using call centres to model the construction process re-engineering (CPR) necessary to make maintenance more profitable.

Conventional wisdom has promoted planned preventive maintenance (PPM) as superior to responsive maintenance, reflecting the maxim 'fail to plan; plan to fail'. However, research has identified that in the commercial sector PPM has been replaced by just-in-time maintenance where an intervention is made just before a component breakdown affects the operation of the organisation (Smyth & Wood, 1995).

Further research (Wood, 1997, 1998a) has shown that the residential market has been targeted by non-traditional suppliers of building maintenance services, notably by motoring and insurance organisations, offering responsive maintenance services along the lines of 'just in time'. At the same time, the UK government, supported by professional and trade institutions, has expressed renewed concern about the poor image of the construction industry created by so called 'cowboy builders' (DETR, 1998). This suggests that builders need to refocus their attention on the needs of the maintenance market if they are to meet client expectations of quality service.

Structure of work

Over the last few decades the structure of work, the workforce, workplace and the working week have changed a great deal. Many of the former labour-intensive heavy industries of the Midlands and north of England, South Wales and central Scotland, including coalmining, steel, shipbuilding and manufacturing, have declined severely, putting many out of work. In large part manual jobs have been replaced by positions in the 'softer' service sector, sometimes in those same regions, often elsewhere, particularly in south-east England.

There are few women in construction in the UK. Latham reported (1994, p.71) that 'women are seriously under-represented in the industry, and the traditional excuses offered in respect of site operatives are becoming less relevant as the building process becomes more mechanised, there is more off-site prefabrication and plant replaces labour'. A study by the author (Wood, 1999a) identified that over 40% of construction firms in Oxfordshire had no female employees and 97% of the remainder had four or less. There are also labour shortages in the region; 40% of Oxfordshire construction organisations said they had had difficulties filling vacancies. It has been difficult for several years for workers from other parts of the country to find affordable housing in the county. These are powerful drivers to rethinking building maintenance services.

The studies presented in the Oxford and Montreal papers (Wood & Smyth, 1996; Wood, 1997) showed that the building maintenance market had been penetrated by many non-traditional suppliers of building-related services. These studies suggested that re-engineering would be necessary for the construction industry to boost its performance to regain competitive advantage.

Service development

The penetration of the building maintenance market by such organisations as Green Flag and the Automobile Association has been facilitated by the provision of call-centred maintenance (Wood, 1998a). These organisations are 'both used to the idea of building and maintaining membership and thereby repeat business ... [and] are therefore very conscious of the need to meet the needs of their members'.

Features of these services include:

- service culture and infrastructure
- substantial customer base
- membership/subscription
- brand identity and loyalty
- reliability and trust
- trained personnel
- care and quality 24 hours a day
- vetting of contractors
- certified skills
- business skills
- communications infrastructure
- regular inspections.

Re-engineering

The aim of re-engineering is to boost performance and improve competitive advantage. Re-engineering is necessary for the construction industry and in particular the maintenance sector of the industry.

According to the principal proponents of re-engineering, Hammer & Champy (1993), 'It [re-engineering] isn't about fixing anything ... [it] means starting all over, starting from scratch'.

Grint & Case (1998), in examining the 'violent rhetoric' of re-engineering, quote this claim for the approach:

> 'drive down the time it takes to develop and deliver new products, dramatically reduce inventory and manufacturing time, slash the cost of quality and win back market share. The following changes are possible: 30–35 per cent reduction in the cost of sales; 75–80 per cent reduction in delivery time; 60–80 per cent reduction in inventories; 65–70 per cent reduction in the cost of quality; and unpredictable but substantial increase in market share.' (Ligus, 1993, p.58)

In essence, re-engineering is about revisiting, re-evaluating and redesigning every aspect of the business, perhaps savagely and radically. There are echoes of this approach in the reports of Latham (1994) and Egan (1998) which recommend radical reappraisal and refocusing of the UK construction industry. Implications of these reports are considered in more detail elsewhere; their application to the maintenance business and maintenance processes was developed in a paper (Wood, 1999c) for the Construction Process Re-engineering Conference hosted by the University of New South Wales in Sydney.

To use re-engineering successfully as a tool for gaining competitive advantage, it is imperative that the process proceeds logically from a knowledge of the current processes, through a holistic detailed examination of realistic options to the choice of an agile maintenance system that can profitably appeal to the market. Although many strategies are possible, the research reviewed the development of maintenance, its management and technology development to show how call centres may contribute as a part of re-engineering maintenance to develop an increasingly 'intelligent' approach to maintenance.

Planned preventive maintenance

The Smyth & Wood JIT study (1995) had reported in relation to supermarkets:

> 'The electronic connection of the checkouts direct to headquarters ... enable information – about the building and its fittings – to be received centrally or transmitted remotely. It is thus possible for central management or a contractor to know that the temperature of a particular refrigerated cabinet at a particular store is rising. By combining this information with parameters derived from commonly collected and assessed performance data, it is possible to determine when an unacceptable situation may be reached, and remedial JIT action ordered.'

This enabled JIT maintenance to be identified and defined as: 'Getting the maximum life from each (building) component and piece of equipment, leaving repair or replacement until the component is broken or fails to function, yet taking action prior to it having a serious effect on the performance of the organisation'.

At the same time, as part of the construction industry is developing with increasing technology and management responses to a changing business environment, much of the industry is populated by unskilled or poorly skilled personnel, unaware or uninterested in such developments. This would be of little concern if it were not for the pervasive negative impact on the image of the construction industry as 'cowboys' or 'builders from hell'.

Predictive maintenance

In referring to the electronic connections between supermarkets and central locations, Smyth & Wood (1995) recognised that data from stores could be combined with other data to inform or determine appropriate JIT intervention. Systems are in operation that bring together performance data in a range of tables and charts that can show trends and relationships.

It was later identified (Wood, 1999b) that:

> 'Control needs more than data; it needs to be coupled with experience, another facet of intelligence, which may be human or artificial. When a particular situation pertains (like an increasing pressure in a pipe) and . . . a particular consequence is likely (e.g. damage) . . . an appropriate intervention (e.g. opening a relief valve) may avoid loss or reduction of service. This is predictive maintenance.'

However, although McCullough (1998) identified benefits of predictive maintenance in terms of increased system availability and cost reduction, Wood & Smyth also identified 'lack of awareness and training to use building management systems (BMS) and of time to look at and evaluate printout. In essence there was as one respondent observed "lots of data; lack of information"'.

That is an interesting contrast to another common situation, that of little data, or of data distributed across a number of participants in the building care and maintenance processes and held 'in their heads'. This is generally recognised as an unsatisfactory arrangement.

> 'As a symbol of its world-class status, the University of Tartu [in Estonia; established 1623] planned to build a new biomedicum to accommodate research and studies in the natural sciences. The university secured a loan from the World Bank to build the facility, on the condition that the university acquire a computer integrated facilities management (CIFM) system to guarantee the successful functioning of the building.' (Archibus, 2002)

Intelligent buildings

A number of authors (e.g. Atkin, 1988; DEGW & Teknibank, 1992; McClelland, 1998) have examined the development of the so-called intelligent building. These portray generally a 'high-tech' approach, giving rise to more or less automated buildings. However, at the 1998 Intelligent Building Conference organised jointly by the BRE and the European Intelligent Building Group (EIBG), a significant strand of thought was that the truly intelligent building was a 'green' building, one that would take the fullest account of passive systems, trying to live with the environment rather than to control it.

Several papers at the conference (BRE/EIBG, 1998) identified conflicts in meeting users' needs for both comfort and control. Some of the presentations referred to instances where automatic systems needed to be overridden by users. Just because control technology exists, that is insufficient reason for its installation and inappropriate systems may give rise to disuse or misuse. Thus although there may be advantage in, for instance, automatic light-switching by presence/absence detection, people may be concerned if it feels like 'Big Brother is watching'.

Intelligent maintenance

The author's paper to the conference (Wood, 1998b) referred to intelligent maintenance as perhaps seeming 'oxymoronic, a juxtaposition of something clean, sophisticated, cerebral with something dirty, inconvenient and carried out by someone in a boiler suit and with an oily rag, eventually'.

A dictionary definition of intelligence quoted in that paper included 'ability to understand, reason and perceive; quickness in learning; mental alertness; ability to grasp relationships; information' (Garmonsway, 1969). Thus intelligence may be located in people, in buildings and in systems. The development of hardware and software systems, including sensors of various kinds, controls, hard-wired, fibreoptic and wire-less systems, has facilitated both the automated/intelligent office building, with fine-tuned, individual controlled environments, remote sensing and control, and the intelligent home.

These developing re-engineered markets are generating demand for differently focused maintenance services, and the serving of this market by 'non-traditional suppliers' has already been referred to. Are there facets of the construction industry that tend toward conservatism and slowness to change while other industries may be more responsive?

Customer care and service culture

These terms are not perhaps commonly associated with construction. This is not to say that this is all the fault of building contractors; the focus of clients is also often on cost and price, particularly on lowest tender, with little attention to time or quality dimensions. 'In general, the construction industry has an overdeveloped sense of cost and an underdeveloped sense of value' (Construction Industry Board, 1998, p.23).

Construction is commonly associated with 'extras', claims, delays, overruns of cost and time, defects, disputes, litigation, low profitability and low repeat business. Often the main 'winners' from construction projects are the lawyers. Increasingly, however, clients, particularly corporate clients with large purchasing power and repeat business opportunities, are exercising their 'muscle' with building contractors and suppliers of maintenance services, in demanding higher levels of service.

Clients are in a position to demand predictable outcomes in terms of time, cost and quality, with delivery on time and to budget, with zero defects, no rework and right first time. The examinations of the UK construction industry by Latham (1994) and Egan (1998) have given fresh impetus to clients and contractors alike to demand and expect higher standards of care and service.

Call centres

At least on the domestic scale, the call centre configuration may offer clients a familiar and perhaps comfortable area of opportunity to investigate.

'Good evening ... how can I help you?' This is putting focus on the customer. Call centres have experienced huge growth in the UK and around the world. An estimate in 1998 (Straum) was that 'there are now 1,300 call centres in the UK'. A further article reported (Groom, 1998) that 'estimates indicate that they [call centres] employ between 150,000 and 320,000 people. Some 75% of companies now have dedicated telephone staff, and investment in telephone contact in sales, marketing and customer service has reached 15.4 billion pounds sterling a year'.

Call centres are characterised by prompt, courteous and scripted responses. Service is normally provided 24 hours a day, 365 days a year, by personnel on short-term and variable contracts. It has been suggested that call centres may be the new 'sweatshops'.

> '... they provide a vital source of jobs in the UK's unemployment blackspots. Data Monitor [a market research organisation] has estimated that 2.2% of the workforce could be answering telephones for a living by 2001 ... Call centre workers, most of whom are women, are constantly monitored and many are on performance contracts. The stress of the job leads to burn-out after an average of about 18 months.' (Denny, 1998)

Some clients may have reservations about such employment regimes when outsourcing.

In relation to building maintenance, in essence the call centre provides the 'shopfront', a single point of contact for the customer, linking to a service infrastructure of subcontractors. All the customer needs to hear after relating the problem in hand is a reassuring 'Someone will be with you within ... minutes'. More will be said about call centres in relation to the provision of building maintenance and care services in Chapter 6.

Construction business process

The construction business in the UK at present is hamstrung by skill shortages. Although the structure of work generally has changed much over the last 20 years, the construction industry has not changed so much. There are challenges to marketing construction-related services in the broadest sense to meet changing client needs and demands and to compete effectively with service providers infiltrating from other industries.

Skills shortages

A report by the Construction Industry Training Board (CITB, 1999) has identified that 'taking into account the need to replace workers who retire or leave the industry, around 70,000 new recruits will be required each year between 1999 and 2003.' ... 'For Building Trades, an increase in formal training of around 5,000 per year (or 25% of the current figure) would be required to meet demand'.

Not only are there numerical shortages. A study by the author (Wood, 1999a) identified that there are also deficiencies in the quality of skills required, including a need for multiskilling to service the maintenance market. At the same time, there is resistance to skills certification and licensing schemes. If the image of the construction industry is to be improved the industry needs to be more attractive to high-quality applicants, to value qualifications more highly and thereby to raise the barriers to entry and thus exclude the 'cowboys'.

Without embracing the potential pains of re-engineering and facing the difficulties that this will bring, the industry must face the possibility of a continuing contraction. For example, in the absence of reliable certification of construction operatives or companies, both AA and Green Flag instituted their own schemes for vetting contractors. The combination of inspections of contractors' premises, insurances and stocks and of completed jobs has given assurance of quality of service. However, this 'overhead' cost is quite substantial and it is perhaps unsurprising that the AA, for instance, moved from a 'select list' of 2500 firms (of 20 000 inspected) that 'either met the AA standards or were prepared to work to them' to about 500 'preferred suppliers'. The AA and Green Flag schemes are discussed in more detail in Chapter 6.

Barriers to entry

There are few barriers to entry to the construction industry; typically all that is required may be an old van, a few tools and a pager or mobile phone. Even at a higher level of operation, barriers are few.

Porter (1980) identified seven generic barriers to entry to a market:

- capital requirements
- economics of scale
- access to distribution channels
- service differentiation
- switching costs
- cost disadvantages independent of scale
- government policy.

Low barriers to entry have allowed others than experienced or qualified construction contractors to enter the building maintenance market. Some of these 'new entrants' are also re-configuring or re-engineering the provision of building maintenance services.

Client focus

Latham and others have identified factors that will need more attention from the construction industry if it is to meet client needs better, including:

- clear focus on the customer
- known delivery time and duration
- care, including tidiness, cleaning up, respect for client's premises, privacy and all that the client holds precious
- quality
- continuity (clients do not wish to constantly 'shop around')
- known price
- no hassles.

The call centre paradigm provides opportunities to attend to these needs.

Re-engineering ways ahead

The studies on which this chapter is based have identified some factors leading toward the development of call-centred maintenance, particularly in the UK domestic market. Observations from the commercial property market also demonstrate the critical need for a human face or interface with information and intelligent systems.

The collection and analysis of maintenance requests via the call centre will also allow the assembly of performance data about particular pieces of equipment and building components in use, from which decisions about component life and replacement strategies may be determined.

Service level agreements may also be entered into with greater certainty and refined over time as confidence is gained about performance achieved against that which was anticipated or hoped for. Such information and intelligence also offer hope that money wasted on building defects may also, over time, be reduced. In 1996, £1 billion of the total £56 billion spent in UK construction was attributed to dealing with defects and Egan estimates that perhaps as much as 30% of construction costs may be attributable to rework.

Remote measurement and monitoring devices allow the development of more automatic maintenance responses, delivered either through the human interface of the call centre or help desk or via electronic control systems as these gain greater acceptability with increasing sensitivity to individual needs.

The call-centred approach could be developed to allow access to a centrally provided helpline, with professional advice and judgement available when required.

The call centre can provide:

- single point of contact
- 24 hour service, 365 days a year

- a friendly voice (and possibly face)
- reliable service
- clean, tidy, careful, high-quality work
- complete service
- surprise-free service.

The call centre also supports the development of the 'learning organisation'. The data collected centrally will provide a substantial database of maintenance requests and work initiated that may help inform future actions and also feed back into design and construction. Participants are thus encouraged to become 'reflective practitioners' and lifelong learners. Re-engineering should be seen as a process rather than an objective and the call centre approach could provide data for further improvements.

Summary

This chapter has interrogated building maintenance under the rather fierce spotlight of 're-engineering'. Fundamental questions have been asked, leading to possible reconstructions of problems as previously perceived. The next chapter applies a view from another perspective, turning the focus from processes to people. The chapters that follow apply that customer focus in more detail and promote a fresh look at the kinds of buildings and services we need for the 21st century.

References

Archibus (2002) Surpassing the standards at the University of Tartu. *Archibus Asset* **11** (1), 8–9.

Armstrong, M. (1994) *How to Be an Even Better Manager*, 4th edn. Kogan Page, London.

Atkin, B. (ed.) (1988) *Intelligent Buildings.* Kogan Page, London.

Building Research Establishment (BRE) and European Intelligent Building Group (EIBG) (1998) *Intelligent Buildings: Realising the Benefits.* BRE, Watford.

Construction Industry Board (CIB) (1998) *Strategic Review of Construction Skills Training.* Thomas Telford, London.

Construction Industry Training Board (CITB) (1999) *Construction Employment and Training Forecast 1999–2003.* CITB, King's Lynn.

DEGW (London) & Teknibank (Milan) (1992) *The Intelligent Building in Europe.* British Council for Offices, College of Estate Management, Reading.

Denny, C. (1998) Remote control of the High Street. *The Guardian*, 2 June, p.17.

Department of the Environment, Transport and the Regions (1998) *Combating Cowboy Builders: A Consultation Paper.* DETR, London.

Egan, J. (1998) *Rethinking Construction.* DETR, London.

Garmonsway, G.N. (1969) *The Penguin English Dictionary.* Penguin, Harmondsworth, p.391.

Grint, K. & Case, P. (1998) The violent rhetoric of re-engineering: management consultancy on the offensive. *Journal of Management Studies* **35** (5), 557.

Groom, B. (1998) Busy line to society's future. *Financial Times,* **21 October**, p.17.

Hammer, M. (1990) Re-engineering work: don't automate, obliterate. *Harvard Business Review* **68**(4), 104–112.

Hammer, M. & Champy, J. (1993) *Re-engineering the Corporation: A Manifesto for Business Resolution.* Nicholas Brearley, London.

Kolb, D.A. (1984) *Experiential Learning: Experience as the Source of Learning and Development.* Prentice-Hall, London.

Latham, M. (1994) *Constructing the Team.* HMSO. London.

Ligus, R.G. (1993) Methods to help re-engineer your company for improved agility. *Industrial Engineering* **25**(1), 58.

McClelland, S. (ed.) (1998) *Intelligent Buildings. An IFS Executive Briefing.* IFS Publications, Bedford.

McCullough, T. (1998) Monitors predict commercial building maintenance. *Business First-Columbus* **14** (31), 34.

McGregor, W. (1994) Designing a 'learning building'. *Facilities* **12** (3), 9–13.

Norton, R. & Smith, C. (1998) *Understanding Management Gurus in a Week.* Hodder and Stoughton, London.

Porter, M.E. (1980) *Competitive Strategy: Techniques for Analysing Industries and Competitors.* Free Press, New York.

Schön, D.A. (1987) *Educating the Reflective Practitioner.* Jossey-Bass, London.

SEEDA (2002) *Framework for Employment and Skills Action.* South East of England Economic Development Agency, Guildford.

Senge, P.M. (1990) *The Fifth Discipline: The Art and Practice of the Learning Organisation.* Doubleday, New York.

Smyth, H.J. & Wood, B.R. (1995) *Just in Time Maintenance.* Proceedings of COBRA '95: RICS Construction and Building Research Conference, Edinburgh. RICS, London, Vol. 2, pp.115–122.

Straum, P. (1998) Property not answering. *Estates Gazette* **9774**, 538.

Toffler, A. (1970) *Future Shock.* Bodley Head, London.

Wood, B.R. (1997) *Building Maintenance Service Procurement.* Proceedings of Conseil Internationale du Batiment Working Commission W92 Symposium, Montreal, pp. 801–811.

Wood, B.R. (1998a) *Maintenance Service Development.* Proceedings of COBRA '98: RICS Construction and Building Research Conference, Oxford. RICS, London, Vol. 2, pp. 169–177.

Wood, B.R. (1998b) Intelligent building maintenance. In: *Intelligent Buildings: Realising the Benefits.* Building Research Establishment, Watford.

Wood, B.R. (1999a) *Construction Skills Training in Oxfordshire.* Centre for Construction Management, Oxford.

Wood, B.R. (1999b) Intelligent building care. *Facilities* **17** (5/6), 189–194.

Wood, B.R. (1999c) *Call Centred Maintenance: Reengineering Building Care Services.* Proceedings of 2nd International Conference on Construction Process Reengineering, CPR 99, University of New South Wales, Sydney, 12–13 July, pp. 131–140.

Wood, B.R. & Smyth, H.J. (1996) *Construction Market Entry and Development. The Case of Just in Time Maintenance.* Proceedings of 2nd National Construction Marketing Conference, Oxford. pp.17–23.

6 Enter Customer Care, Contact and Call Centres

The previous chapter focused somewhat on issues related to process and application of a rather 'hard' approach, looking at maintenance and management methods afresh, being prepared to question and jettison familiar ways. That approach and the 'answers' that arise can be very challenging and hard for some to respond to with enthusiasm. By contrast, this chapter takes a 'soft' focus, putting people, and the customer in particular, at the centre of considerations. These thoughts, which informed the preparation of a paper for a conference organised by the International Council for Research and Innovation in Building and Construction (CIB) in Cape Town, South Africa, represent a transition to a more caring approach.

Background

Earlier chapters have identified how building maintenance has been regarded as a lowly, unattractive activity, despite the large size of the maintenance market. Furthermore, the focus of building research and development is very substantially directed towards the new-build sector, although this may be changing. 'For instance, in the commercial sectors we have seen the growth from nothing to an estimated $1300 bn market for the discipline known as facilities management' (Lowe, 1996).

The facilities management (FM) discipline recognises the value of a building or stock of buildings to the effective and efficient carrying out of an organisation's business. Inadequate or poorly maintained buildings may result in losses of production or productive capability. There may be direct losses due to unavailability of buildings or parts thereof or the services within, and there may be losses due to the effect on staff morale and motivation. Such losses may be qualitative and/or quantitative.

At the same time, technology and management techniques have been developing in ways which are enabling maintenance services to be reconfigured. Studies by the author have codified a number of recent developments, including, for instance:

- just-in-time maintenance (Smyth & Wood, 1995) (see Chapter 4)
- call-centred maintenance (Wood, 1999c) (this chapter)
- intelligent building maintenance (Wood, 1998) (see Chapter 7).

More is said on these elsewhere; suffice it to say here that common features include:

- technological 'connection'
- responsiveness to need
- a service culture
- customer focus.

These approaches have challenged the 'received wisdom' of planned preventive maintenance programmes and are promoting a more thorough review of practice and the underlying rationale for interventions.

Supermarket lessons

The supermarket study that demonstrated the JIT maintenance approach (described in Chapter 4) showed how the stores were applying priorities and methods from one part of their operation to another. This internal exchange of institutional learning provided fresh insights and challenges to previous practices. The same electronic connections between the checkouts and headquarters that enabled information to be transmitted about product sales were also enabling temperatures of refrigerated cabinets to be monitored and remedial attention activated remotely, and perhaps automatically. This 'technology transfer' was also mirrored, and its value enhanced, by the parallel transfer of management techniques. One of the contributors to the economic success of the UK supermarket chains has been their supply chain management – they have built strong long-term relationships with suppliers, based on high quality and reliability. The study identified that the client was increasingly expecting its maintenance service providers to adopt a similar approach, and procurement systems were adapted accordingly.

Existing providers of maintenance service to the supermarkets were engaged in discussion with the client to determine what might be reasonably attainable performance standards, response times and related unit rates. One-year term contracts were drawn up, with intermediate break or renegotiation points to protect both parties in the event of serious misunderstanding or miscalculation, and with options to extend for a further period. Over time, service providers have been able to offer, with confidence, quicker response and cheaper rates, with associated longer term contracts.

Customer care

The supermarket study has shown something of the scope for developing beneficial long-term relationships between clients and contractors, as trust is built and reliability achieved. This is quite a contrast with the general image of the construction industry, which is often characterised by:

- late delivery
- extra costs
- defects and low quality
- poor skills
- adversarial relationships.

It is common for builders, particularly small, general builders, to concentrate on the issues of price and managing a fluctuating workload, at the expense of focusing on customer care. Smaller builders, and those involved in maintenance especially, also tend to comprise personnel with few, if any, qualifications. Those qualifications are also more likely to be craft related than management or professionally orientated. A City & Guilds Certificate is perhaps the most likely and a Higher National Certificate (achieved by two years' day-release study) is the highest normally found. Degrees are still relatively rare in this field. Many in the industry have had no education or training in business skills (Wood, 1999a). Dissatisfaction with small domestic works particularly, and including repair and maintenance work, spurred the UK government to launch the 'Combating Cowboy Builders' initiative referred to elsewhere. The 'Cowboy Builders' task group looked at examples abroad, including Australia (Georgiou *et al.*, 1998), and their proposals included:

- encouragement of the provision of 'approved builders lists' by local authorities
- development of the Construction Skills Certification Scheme (CSCS) for use by domestic customers
- extension of the Building Regulations
- greater involvement by lenders and insurers
- ABTA – style bonding (ABTA – the Association of British Travel Agents – has a bonding scheme to fly home passengers stranded by the insolvency of their carrier).

The CSCS is administered by the Construction Industry Training Board (CITB) and is financed through its continuing levy on construction firms. In essence, certified individuals carry colour-coded cards that indicate their level of skill or competence. It has been noted already, however, that many of the construction qualifications, including National Vocational Qualifications (NVQ), do not cover business skills and it has been recognised that this deficiency should be redressed. It has also been recommended that 'approved' or 'registered' or 'accredited' building firms should be required to maintain a minimum proportion of suitably qualified personnel. In the meantime, the lack of such registration or regulation in the UK has encouraged at least two organisations, the Automobile Association (AA) and Green Flag Insurance to offer building maintenance services using vetted contractors. Although the AA has subsequently withdrawn from this market, the Direct Line insurance group has recently added its 'Response 24' service. These services are examined in more detail below.

With regard to the other 'cowboy' constraints proposed above:

- individual local authority 'approved lists' are being discouraged in favour of one national list compiled and maintained on behalf of central government and marketed as 'Constructionline'
- the application of Building Regulations has been extended to some work previously exempt from control as repair work. For instance, when windows are replaced they must now conform to the standards of a new window, i.e. provide much better thermal insulation than an old one; generally this means upgrading to double- or triple-glazed units. A formal application is also required unless the installer is registered under the FENSA scheme. FENSA (Fenestration Self-Assessment) is the scheme developed by the glazing industry to allow the self-certification of replacement windows under the Building Regulations by installers. FENSA is owned by the Glass and Glazing Federation (GGF). Details of the scheme can be found at www.safety.odpm.gov.uk/bregs/news/fensa/index.htm
- more insurers are requiring that work is only to be executed by contractors that are on their own approved list. Often householders are only too happy to do this, as it provides them with some security that the work should be carried out satisfactorily
- bonding is being less pursued. In furtherance of Latham- and Egan-inspired initiatives, the deduction of retention sums from interim payments through contractual arrangements is being frowned upon, being replaced by a climate of trust between the parties.

Building care

Building care represents a paradigm shift in the understanding and practice of maintenance. A definition might be:

> Building care is the pursuit of the enduring supply of the best environmental conditions in which to support the corporate objectives of the organisation.

Features of building care include:

- intelligent use of technology
- 'light touch' management
- responsiveness
- control generally at the individual level
- congruence with corporate strategic directions.

Naturally there will be difficulties in attempting to meet potentially conflicting demands of individuals and 'the corporate body' and in balancing needs and wants. Essential first steps (which would need to be continuously revisited, reviewed and revised) would include identifying those wants and needs. Comfort and control are important and terms like 'best practice', 'empowerment' and 'realising potential' are appropriate considerations. Perhaps the concept of 'care' demands a softer rhetoric than that associated with the 'oily rag' approach to maintenance; redescribing, relearning and refocusing rather than re-engineering. The experiential and reflective styles of Kolb (1984), Schön (1987) and Senge (1990) referred to in Chapter 5 are very relevant here.

Users, care and service

A number of facets and factors conducive to building care can be identified.

- Building users are looking for service.
- The 'conventional' construction industry has been losing markets to 'non-traditional' suppliers with service cultures, such as catering, cleaning, security and motoring organisations.
- The market for reliable customer-focused maintenance services is substantial; the construction industry is responding slowly to market needs.
- Maintenance providers have been re-engineering service in response to user demands.
- There are models of customer care from which the building maintenance industry can learn.
- Technology is available to assist effective responsive maintenance; intelligent application of such technology is required.
- The construction industry has a poor record of learning; personnel are poorly qualified; business skills are often lacking; closer links between construction and education are needed.
- Although maintenance is not 'sexy', there is scope for redefinition and refocusing with a developing concept of 'building care'.

A wider view

Earlier chapters have shown how the 'orthodoxy' of PPM has been challenged by the application of new techniques and theories such as JIT. The growth in services technologies, through computers and hard-wired systems to telecommunications and remote sensing devices, has enabled the so-called intelligent building. This has also facilitated further development in maintenance service, in a direction that is captured in the term 'intelligent building maintenance'; this development is discussed in detail in Chapter 7. The technology has also made possible the provision of service in a more 'personal' way, enabling customer care to be delivered. However, as the building industry has not been noted for its attention to customer needs it may not be a surprise that this has been taken up by organisations from outside construction, 'trusted organisations like the Automobile Association and Green Flag Insurance using their communications infrastructure linked to certified, trained personnel and guaranteed delivery' (Wood, 1997).

The AA and Green Flag initiatives were focused on the UK residential sector, where much had changed. Largely through the government-promoted 'right to buy' initiative of the 1980s, local authority-owned and managed housing declined from around a half to about a quarter of the housing stock (DETR, 1997). This change created a very substantial market of clients for building maintenance, repair and improvement services, relatively inexperienced and uneducated for the task. The opportunity to serve this market has been seen by AA and Green Flag who are competing against a still largely unco-ordinated, customer-unfriendly and fragmented construction industry. This maintenance service development is a major component of a holistic, customer-focused approach to building care.

The housing maintenance market

The configuration of the UK construction industry and its impact or lack of it on housing maintenance were described in Chapter 4. There is still substantial work to be done. The English House Condition Survey carried out in 1996 and published in 1998 (DETR, 1998b) was referred to in Chapter 1. It showed continuing disrepair and unfitness in the housing stock.

- Nearly 80% of dwellings had some fault recorded to the external or internal fabric.
- The average level of disrepair across the whole stock was about £1500 for an average size house.
- The highest levels of disrepair were found in older, inner London boroughs, large urban districts and older resort and university towns.
- The mean cost of repairs and replacements due over the next ten years for owner-occupied semis and terraces built pre-1919 was £3460 and for private rented pre-1919 houses was £5320.
- There were 1 522 000 unfit dwellings, some 7.5% of the stock. About 1 million of these dwellings were also recorded as unfit in 1991 but the remaining half a million became unfit since 1991.
- The most common reasons for unfitness are unsatisfactory facilities for preparation and cooking of food, disrepair and dampness.

This combination of size and scope of need and the way that the market was being served, or not, by the construction industry provided the opportunity for 'non-traditional' players to enter this market.

But why would a concern, such as a motoring organisation, want to get involved with home repair? As Cahill & Kirkman reported (1994) on the development of a home emergency service by National Breakdown (now Green Flag Home Assistance Services):

> 'National Breakdown is well-known as a motoring organisation and has 2.5 million subscribers'. '... over the last 21 years a service has developed which relies on telephone communication skills and in managing sub-contracted workforces. All the calls are centrally administered by trained people with telephone-based diagnostic skills, who farm jobs out anywhere in the country to skilled and vetted people working on National Breakdown's behalf.'

The knowledge and skills developed by the motoring organisation enabled it to expand into home emergency services. Features of this 'call-centred maintenance' service are described after an introduction to the features of the call centre.

The call centre or contact centre

A call centre is a hub for dealing with communications between customers or potential customers and an organisation. Generally those communications will be by telephone, although over recent years email has become increasingly significant. The call centre may be a part of the organisation which it serves, either co-located or elsewhere, or it may be provided as an outsourced function. A call centre is typified by large office floors populated by many anonymous identical workstations or cubicles, each with its own telephone and computer terminal, each occupied by a 'customer service advisor' or the like. These 'centres' are often located in large industrial warehouse type buildings on 'business parks' on the suburban peripheries or as part of a 'regeneration' of former industrial inner-city areas. The buildings and locations are often selected for cheapness and proximity to a readily available workforce. That workforce need not be well qualified, and therefore likely to seek high wage rates, but susceptible to training in scripted 'customer care' routines. Indeed, it is possible to provide the call centre function as a 'virtual' or 'dispersed' centre through people working at home. This enables people to work what may otherwise be considered 'antisocial hours' without concerns about travel and related expense or personal safety issues. These workers are often women with childcare responsibilities and few other work opportunities, concentrated in rundown inner-city or peripheral housing estates and keen to increase their earnings however marginally above Social Security benefit levels.

More recently, call centres have been relabelled as 'contact centres' which is perhaps intended to convey a sense of fuller engagement than a mere taking of calls.

Features of the call centre

'Good morning (afternoon/evening) Mr (Mrs/Miss/Ms) Bloggs (Joe/Jo), Chris speaking, how can I help you?' This is the start of a conversation, a potential relationship in the making. What follows is aimed at identifying and satisfying a customer need. In the first few words the customer is to be put at ease, confident in the knowledge that they and their problem are being taken care of. This is facilitated in a number of ways.

- The telephone number will be chosen to be easy to recall from memory; it may be provided also on sticky labels to put prominently by the phone lest we forget.
- The call will be answered promptly. Typically calls will be expected to be answered within a prescribed number of rings. It is frustrating to wait what seems like a long time before someone answers. As time stretches out, we may think that we will be passed to a less than satisfactory answering machine or voicemail system. Systems designed to lead the caller to the most appropriate 'service' by pressing the related 'touchtone' button as directed may be cost effective (i.e. cheap) for service providers, but personal experience tells us that we want to speak to and be listened to by a real person. We do not want to be put 'on hold', even if accompanied by Vivaldi or some other 'soothing' muzak. 'Your call is important to us' sounds hollow when your call is held in a queue. This means that service providers need to understand and anticipate call patterns, such that their staff and systems are not overwhelmed at times of peak demand, which may vary from day to day. The length of time to answer, the start of the dialogue, may be seen as an index of efficiency, a performance indicator.
- Calls are normally taken over a period longer than what used to be called the working day, typically 9 a.m. to 5 p.m., Monday to Friday. Many commercial enterprises and most residential clients require service 24 hours a day, seven days a week. Indeed, it is the 'outside normal hours' that is most needed; many emergencies will just not wait until Monday morning for attention.
- The call will be answered with courtesy and consideration. The caller, a member or subscriber or potentially so, may be in a state of distress or exasperation; they may be frustrated and at their wit's end. They may be angry, especially if they have waited a long time to get through. It is the job of the person taking the call to exude calm. The operator may be assisted in this task in a number of ways, such as:

  - having an acceptable accent
  - induction and training
  - software that identifies suitable solutions from keywords

- a scripted response appropriate to the situation
- appropriate periods of rest and recuperation.

Perhaps the 'acceptable accent' may seem a peculiar point, but apparently market research has shown that people respond more favourably to some accents than others. In the UK, call centres are more common in Belfast, Leeds or Newcastle than in Liverpool or Birmingham (Wood, 1999b).

- The person taking the call is generally expected to deal with the whole of the enquiry or transaction with the caller. The operator will almost certainly have at best only limited practical experience of building construction and maintenance in general and almost certainly no technical knowledge related to the particular building or problem in question. Their response is prompted by predetermined questions displayed in turn on the screen. Answers keyed in will lead to further prompts and eventually a 'solution' will be arrived at. Interrogation of this kind is needed such that an appropriately skilled person can be identified and despatched. Again this is what the customer wants. There is no value in sending someone to the job unable to fix it.

  There may be scope here for growing and deploying a more multiskilled workforce. The construction industry is still tending to produce 'craftsmen' trained in a single building skill, such as bricklaying, carpentry or plumbing, and this is limiting. However, the expression 'jack of all trades, master of none' should be borne in mind.

- The software should thus identify the operative most readily available to respond to the situation and their estimated time of arrival. This too is important to the caller. What is wanted is 'someone (or better perhaps, Chris, a fully qualified electrician, been with us twenty years) will be with you within the hour/will be there at 2.15. Don't forget to check her ID and ask for her password which you have just given me, before you let her in'. The time taken to get to this point is important, but the time from call to attendance is the more critical performance indicator. Knowing that 'someone is on the way' helps ease the pain of waiting and avoids anxiety.

- Remembering the dictum that 'last impressions count', the caller may then be asked 'Is there anything further I can do for you now, Mrs Bloggs?' or perhaps 'I'll phone you around 3.30 to check everything's OK'. (Of course, first impressions also count, but apparently last impressions count more.)

- Although the software may be quite expensive, representing many hours of professional work to identify appropriate questions to translate symptoms into possible causes and likely work required, this investment repays when processing calls through more lowly paid operatives.

Features of call-centred maintenance

Green Flag and the AA are both used to the idea of building and maintaining membership and thereby repeat business. They therefore avoid the need to keep looking for new work for new clients, tendering for that new work and losing five in six tendering competitions. They are also very conscious of the need to meet the needs of their members. This manifests itself in the following features (which are very much in line with the recommendations of Latham for an improved construction industry):

- focus on the customer
- delivery times
- care
- quality
- continuity
- known price.

The customer for these services is often going to be calling in a situation of some distress, perhaps panic, and a professional response that puts the customer as the first concern is welcome. As Cahill & Kirkman identified, with appropriate technical training the assistant can help the caller to cope until the tradesperson arrives, for instance advising on how and where to close a water stop-cock, thus preventing further damage. Service like that gives 'peace of mind'.

This culture of concern and confidence is exemplified further by requirements that all subcontractors carry identity cards and by the provision as standard of a questionnaire sent out after every job to track what was done and check customer satisfaction. Indeed, the whole service is focused on trying to give customers (described as subscribers or members) what they want. Market testing is also done to ascertain how the product may be developed. This is likely to result in newly packaged or positioned products in the future as the AA, Green Flag and others review their portfolios.

Responsiveness and responsibility

Often the customer's main concern is for a prompt response, for someone to attend soon and to have the problem rectified quickly. Sometimes, service is so bad that the customer is relieved to even make contact with someone on the telephone and whose undertaking is to 'get someone there to look at it sometime in the next day or so'. Delivery times may not be as important as just having certainty that someone will come, although as a rule sooner is certainly preferable to later.

Although guaranteed response times were not quoted, promotional literature from the AA and Green Flag indicated that 'contractors were on the premises within 90 minutes on average' and that 'the "standard service" of 3 hours labour absolutely free is ample time to solve most household problems'. It is common in the public sector now to quote response times and for performance to be measured against benchmarks. It would be reasonable to expect that with experience and confidence, norms could be developed. The new entrants have recognised the need to meet customers' expectations in terms of quality of service or process, which I am calling care, and the quality of the end-product, the repaired item. The questionnaire has already been mentioned.

Reference has also been made to home owners' concerns about 'inappropriate behaviour'. We do not wish, particularly when feeling vulnerable, to have ourselves, our homes and possessions abused, our coffee taken (unless we have offered), and the vetting of these attitudes and behaviours of tradespeople is perhaps as important as their technical skills.

The lack of technical skills was the principal concern of the DETR 'cowboy builder' consultation. 'The biggest problem facing the domestic consumer when selecting a builder lies in distinguishing between builders who can be relied upon to carry out a good job to a fair price and those likely to supply a defective service, in all probability over-charging in the process' (DETR, 1998a).

Assuring standards

The DETR did not consider compulsory registration for builders to be feasible for reasons of practicality, cost and enforcement, but offered a number of alternative proposals. Although more will be said later in this chapter on responses to the DETR consultation, suffice it to say at this stage that the absence of such provisions led the AA and Green Flag to institute their own schemes for vetting contractors. Green Flag, for instance, rejected all firms of less than six people as being unable always to respond immediately to an emergency and insisted that all people within the firm had their own communication links (mobile phone, pager, etc.). Inspectors were sent to look at completed jobs to check the standard of work and level of service and 'in use' unannounced inspections are made to check on, for instance, insurances and stock levels. For the AA, each company was visited at least once a year, unless there were already concerns, to monitor changes.

The scope and scale of vetting were substantial. As reported earlier, 20 000 firms were inspected, of which 2500 either met the AA standards or were prepared to work to them, and the number in 1998 was down to about 500 as the AA worked towards 'preferred supplier' arrangements. Jobs such as damp proofing or drain cleaning were managed by large national or regional companies whereas more rural and jobbing emergency call-outs may still be handled by smaller firms. Size of firm for reliability or response and quality of work were becoming major concerns.

Finally in this section I turn to considerations of continuity and price. For the contractor or service provider there is continuity of work, with orders (or calls) coming in every minute of the day and night, 365 days a year, which together with scale permits economy. For the subcontractor (the small builder, tradesman, maybe sole trader) there is also a continuous (or semi-continuous) flow of work provided that he or she maintains the standards set by the contractor. The AA was careful to mention that they don't, however, allow any subcontractor to become too dependent on them as a source of work. This continuity of work and a stable contractually defined relationship allows for keen and certain pricing, a benefit which is passed on to the customer.

Whither the cowboy?

Chapter 4 described something of the nature of construction firms operating in the maintenance market: small, poorly qualified, unfamiliar with good management practices. This section looks at how it is that such bad firms survive or even thrive. Factors conducive to the cowboy include:

- uneducated and inexperienced clients
- clients in distress, panic
- poor/few sources of guidance or help
- fragmented industry
- do-it-yourself
- discontinuous workloads
- lack of contract infrastructure.

Typically home owners do not understand the maintenance needs of their properties, do nothing until something goes wrong, then do not know who or where to turn to and are often only too pleased to find someone at last who says they will attend to it soon.

In these circumstances it is very difficult for clients to ask (or even think to ask) about matters such as qualifications, quality, references, timescale and price. If we are to improve the construction industry's performance and the public's perception, we need to address the environment in which the cowboy thrives. This should include initiatives such as:

- development of the school curriculum to include 'looking after your home' and/or providing this through lifelong learning opportunities, e.g. short courses, evening classes at colleges, presenting the benefits of planned, preventive and cyclical approaches, and of using professionals
- encouragement of 'one-stop shop' advice/call centres, perhaps insurance linked, where 'advice' could extend to recommendation or organisation of an appropriate builder or other professional
- development of the Construction Skills Certification Scheme (CSCS), recognising the needs for multiskilled operatives supported by professional management, perhaps along the lines of the Chartered Building Company (CBC) scheme of the Chartered Institute of Building
- promotion of quality, through extension of building regulations, development of national (or international) standards, government-supported advice akin to that on energy efficiency

- development of a simple standard contract that could be 'implied' when not explicitly entered into
- the promotion of more opportunities for women.

Such initiatives correspond well with the government consultation paper mentioned earlier. That paper listed ten proposals ranging from more advice, through more regulation and/or certification to easier redress. As the executive summary puts it (DETR, 1998a), 'Tackling the problems caused by so-called 'cowboy builders' in the repair and maintenance sector will require action on a broad front to improve service to the consumer and raise standards'. In summary, the ten proposals relate to:

- approved lists
- Constructionline
- CSCS
- kitemarking
- extending building regulations
- involving lenders and insurers
- ISO 9000
- warranties
- bonding
- dispute resolution.

Responses to the consultation were broadly supportive and encouraging. There was a general consensus that 'something must be done', though less certainty about what should be done. A number of the proposals were taken forward, including Constructionline and CSCS. Their introduction and take-up have been fitful, but more time is needed to see results. Responses included the following

- The industry must raise its game to provide better, more reliable services and appropriate forms of contract (RICS).
- FMB welcomes the new willingness by the government to tackle this long-running problem (Federation of Master Builders).
- The CIOB supports the need to address the issue of the cowboy builder/rogue trader once and for all.
- We are pleased that the government does not consider it practical to introduce compulsory registration of builders (or nationalisation) but the first proposal could lead to this (FMB).
- Much needs to be done to educate the customer (Construction Industry Council).
- Government, industry and professional bodies must seize this opportunity of raising standards and turning a much maligned industry into one of world-class standards (CIOB).
- CSCS is considered to be of limited scope and value (CIC).

- There is now no excuse for not insisting that management and higher technical level staff in all companies are properly qualified (CIOB).
- FMB believes that any 'kitemarking' scheme must be comprehensive and cover workmanship and also areas such as sound financial management, provision of technical support and commitment to training and skill enhancement (FMB).
- If an industry-wide 'kitemark' is to be credible it must be subject to the highest standards of enforcement. The Master Plumbers' and Mechanical Services Association of Australia expelled or suspended 16% of its members in 1996–7 for failing to come up to the required standards of competence. How many members of British trade associations have membership terminated for unacceptable standards of conduct or competence? (CIC)
- Any repairs should be undertaken to the current standards of the Building Regulations (RICS).
- Lenders show little or no interest in imposing restrictions on the selection of contractors (CIOB).
- Some insurers (e.g. Royal and Sun Alliance) are already signing up for 'block' R & M [repair and maintenance] work for their insured with quality traders in given areas (CIC).
- We would recommend the development of easy-to-use, simply explained technical standards similar to those provided by NHBC (CIOB).
- There are many existing 'guarantee' schemes operated by trade bodies within the industry. The comparative merits of such schemes in terms of customer benefit are not widely known (CIC).
- Any regulatory approach should be seen as a fall back. This would be complementary to education of the public (RICS).
- It will often be difficult to ensure that decisions (dispute resolution, arbitration, etc) are enforceable (CIC).
- The conclusion must be that the most cost-effective solution is prevention (RICS).

One of the motivations for the AA becoming involved in building maintenance was to provide a service to its members from a background of representing the member when dealing with complaints about, perhaps, a second-hand car or the maintenance or servicing of a vehicle. Comparison with the motor trade is informative. Before the introduction of the MOT test, many an unroadworthy vehicle would be driven until it was involved in an accident; at the same time cars were getting faster and motorways being built on which they could be driven fast. It was seen that to 'keep death off the roads' standards for brakes, lights and steering had to be introduced and enforced. It was also common in earlier years that owners would maintain their own cars or they may have had them serviced by unqualified 'mechanics', perhaps friends or neighbours. It is now much more common for cars to be serviced according to predetermined schedules, say every six months or 9000 miles, by authorised dealers with pricing according to published 'menus'. Whilst recognising that there are many differences between homes and cars, it is instructive to see the 'makeover' that has been achieved in another previously maligned and disreputable industry and to see that there are grounds for hope that the image of the builder can be turned around.

Where there is a vacuum, a job to be done, someone will seize the opportunity. Architects, for example, were not very good at the project management role – their education was often more focused on design so others calling themselves project managers, and better qualified in talking about time and money to clients, moved in to perform many of the roles that the architect had previously filled.

In looking after buildings, maintenance engineers and surveyors have lost out to cleaning, security and catering firms with a can-do approach and service infrastructure in the 'new' area of facilities management. The rise of building and construction management degrees has come late. Degrees are not yet a common feature of the maintenance landscape. However, the contractor may be doing little but organising subcontractors, some of whom today are very specialised and professional, while others are selected on little other than price and availability; the concept of partnering is still relatively not trusted. Green Flag and the AA recognised a need and a gap in the market and have been filling it. Whilst there are signs that they may have underpriced or underestimated the complexity of the delivery channels, there is little indication that the construction industry has sought to respond more actively.

Beyond the call centre: coming closer to home through the help desk

In some ways the call centre has brought maintenance closer to the building user; it may be that for much work the need to track down the maintenance manager or building manager or estates manager has been circumvented or cut out completely. However, this has brought with it a certain depersonalisation. An attempt has been made to remedy or at least reduce this shortcoming for occupiers of commercial buildings, especially offices, by the introduction of the help desk.

The help desk is generally staffed at times when the building is normally occupied and may be co-located with the main building reception. Not only will the staff take the telephone calls that would otherwise go to the call centre, and deal with them in similar fashion, but they are on hand to be approached personally. If such staff are seen each day, a real relationship can be built up. A smiling face may help to generate and maintain an impression of true care and concern for the working environment of a friend. It may also be that in the case, for instance, of a serviced office arrangement, the help desk may be a vehicle for the delivery of a seamless, all-inclusive service.

Such a service, although located 'on the premises', may be provided on an outsourced basis by a service company or contractor, allowing the building owner or occupier to concentrate on their core business. The contractor may provide staff and/or equipment and systems and they may operate to all outward appearance as if they were employees. They may be expected to wear an appropriately badged uniform in order to project a desired image. For some service providers, this seamless delivery is an important part of their offer, which may be presented as a total facilities management (TFM) package. TFM is especially attractive to those who want minimal opportunity for 'boundary disputes' between providers of related services. Although such a service may be more expensive, there is 'one bum to kick' when things go wrong.

Conclusion

The picture painted of the building maintenance industry is fairly unattractive. It is populated by small firms, poorly qualified, doing a poor job, poorly organised and poorly remunerated, working hand to mouth and held in low regard. Whilst many of the problems recounted may relate to rogue traders, many are matters more to do with incompetence or ineptitude and susceptible to improvement by education and training. There is also a role for greater regulation or certification and for redress when things go wrong.

The good practice suggested here must be promoted by education, professional bodies and trade associations and taken on board by contractors. The evidence shows that when good service is given customers are attracted and this gives reason to believe that significant improvement in service quality can be achieved. There are features of successful service delivery that might be emulated by firms wishing to enter or to enlarge their share in the building maintenance sector. Although this chapter, and the book generally, does not address questions of cost or whether this is a way to make money in the industry, it has been shown that for many customers price was not the main determinant of contractor selection. Reputation and having worked together before were strong considerations. The evidence suggests that if owners are provided with an infrastructure that allows them to make an informed choice, they are more likely to commission a better builder and to receive a better job. At present many are making uninformed, uneducated choices, paying the price and learning 'on the job'. This is expensive and undesirable; it is also unnecessary.

The UK government, professions and trade associations are 'on the case' and presenting a reasonably uniform understanding of the nature of the problem and ways in which it may be tackled. The studies on which this chapter is based have also shown that when addressed professionally, building maintenance can be organised effectively.

The motoring organisations have a service infrastructure based on serving their members and they are on call at all hours of the day and night and care for their customers. Those taking the calls are trained in interpersonal skills and have appropriate technical back-up. The construction industry, more accustomed to the mess and dirt of the building site, needs to 'smarten up' and project (and deliver) a fully professional service. It is often suggested that 'the construction industry is different' and that it has nothing to learn from other industries. It may be that many in construction have learnt little; others have learnt more and are developing markets at the expense of the traditional 'builder'. Is it too late to learn? This chapter has demonstrated something of what can be learnt; there is hope that the cowboy may soon be on his horse out of town or looking for a management course to improve his customer care.

Summary

This chapter has identified how some maintenance services have been arranged around contact centres and help desks, putting people at the centre of operations. A number of issues relating to the rather neglected role of the customer in achieving well-maintained buildings have been examined. While it may be thought that the professionally perceived technical needs of the building should be paramount, it is suggested that perhaps 'the customer knows best'. This focus on personal priorities and feelings rather than technical fact represents a significant paradigm shift and requires of some organisations and individuals a major change of culture to adopt a new approach. The next chapter looks at how some organisations have responded, and some of the consequential effects on people involved in the processes.

References

Cahill, P. & Kirkman, J. (1994) Home emergency services. In: *Encouraging Housing Maintenance in the Private Sector* (Leather, P. & Mackintosh, S., eds) Occasional Paper 14. SAUS, Bristol.

Department of the Environment, Transport and the Regions (1997) *Housing and Construction Statistics*. DETR, London.

Department of the Environment, Transport and the Regions (1998a) *Combating Cowboy Builders: A Consultation Paper*. DETR, London.

Department of the Environment, Transport and the Regions (1998b) *English House Condition Survey*. DETR, London.

Georgiou, J., Love, P.E.D., Smith, J. (1998) *A Comparative Study of Defects in Houses Constructed by Registered Builders and Owner Builders*. Proceedings of the 32nd Annual Conference of the Australia and New Zealand Architectural Association (ANZASCA), University of Wellington, New Zealand.

Kolb, D.A. (1984) *Experiential Learning: Experience as the Source of Learning and Development*. Prentice-Hall, London.

Lowe, K. (1996) Best of all worlds. *Property Week*, 25 January **54** (4), 38–40.

Schön, D.A. (1987) *Educating the Reflective Practitioner*. Jossey-Bass, London.

Senge, P.M. (1990) *The Fifth Discipline: The Art and Practice of the Learning Organisation*. Doubleday, New York.

Smyth, H.J. & Wood, B.R. (1995) *Just In Time Maintenance.* Proceedings of COBRA '95: RICS Construction and Building Research Conference, Edinburgh. RICS, London, Vol. 2, pp. 115–122.

Wood, B.R. (1997) *Building Maintenance Service Procurement.* Conseil Internationale du Batiment W92 Symposium, University of Montreal, pp. 801–811.

Wood, B.R. (1998) Intelligent building maintenance. In: *Intelligent Buildings: Realising the Benefits.* Building Research Establishment, Watford.

Wood, B.R. (1999a) *Construction Skills Training in Oxfordshire.* Centre for Construction Management, Oxford.

Wood, B.R. (1999b) *Building Care through Customer Care.* Proceedings of Conseil Internationale du Batiment Joint Symposium of Commissions W55 and W65, Customer Satisfaction: a Focus for Research and Practice, University of Cape Town, September, pp. 1069–1077.

Wood, B.R. (1999c) *Call Centred Maintenance: Reengineering Building Care Services.* Proceedings of 2nd International Conference on Construction Process Reengineering, CPR 99, University of New South Wales, Sydney. 12–13 July, pp. 231–140.

7 Intelligent Building Care

For many, the juxtaposition of a word like 'intelligent' alongside 'building' or 'maintenance' will seem like an oxymoron. However, not much thought is needed to see that buildings need intelligence to be applied for their most effective maintenance. Similarly, the so-called intelligent building also needs maintenance which should be carried out intelligently. What does 'intelligence' mean in this context? This chapter identifies possible locations of 'intelligence' in people and buildings. It also examines the relationship between technology and users, particularly regarding user satisfaction, and addresses both the care of intelligent buildings and the intelligent care of buildings. It juxtaposes the 'clean, sophisticated, cerebral' associations of intelligent buildings with something 'dirty, inconvenient and carried out by someone in a boiler suit and with an oily rag, eventually', which is many people's perception of maintenance.

Intelligence and the intelligent building

'Intelligent buildings are structures equipped with computer controls that respond to complex stimuli to increase the buildings' efficiency and profitability' (Sullivan, 1994). This is one of many attempts at definition focusing on automation; a delegate at the 1998 BRE/EIBG conference used the term 'gizmology'. Other definitions suggested at that conference included the following.

> An intelligent building maximises the efficiency of occupants and allows effective management of resources with minimum life-time costs (EIBG).
>
> One that is more responsive to user needs with the ability to adapt to new technologies or changes in organisational structures (DEGW).
>
> An intelligent system is one that has intelligence applied to it during the design and implementation stages, and allows individuals using it in their daily lives to use their intelligence, and think and act for themselves (Steve Woodnutt, Delmatic Lighting Management).
>
> The intelligent building uses the best available concepts, materials, systems and technologies to deliver the most benefits to building stakeholders while using the least resources. In short it uses the best to deliver the most while using the least (Nick Thompson, Principal of Cole Thompson Associates, architects of the Integer House at the BRE site, Garston, UK).

> An intelligent building should incorporate the latest and best systems, technology and design in order to deliver improved building performance. However ... the latest has still to be invented! Flexibility to allow new technology to be 'plugged in' later is a necessary part of the design process (Ross Campbell, Northern Ireland Housing Executive).

A definition of intelligence (Garmonsway, 1969) includes the following:

- ability to understand, reason and perceive
- quickness in learning
- mental alertness
- ability to grasp relationships
- information.

Such intelligence may be sought today in people and in buildings! Until recently this was in short supply in both. Graduate maintenance personnel are unusual; although there are well-qualified building services engineers they are focused on design, installation and commissioning, while for maintenance personnel the highest qualification in the UK is generally a sub-degree level qualification, such as a Higher National Certificate.

Data and information

Similarly, intelligence in the military sense of information has also often been hard to come by. Although there may be extensive records, for instance works requests, orders, invoices, etc., it would be less common to find analysis of time taken from request to completion or frequency of call-outs, repeat work, patterns of failure, etc.

Technology advances have now enabled firstly the processing and analysis of such data and subsequently the collection and production of even more data than could ever have been envisaged, imagined or used. It is here that the technology of intelligent buildings enables them to be cared for more effectively, without the need for human intervention at the point of need to identify what the data means.

For instance:

> '... today's state-of-the-art automated air-conditioning systems go far beyond monitoring ... to indicate whether the system is simply on or off. Rather, they regularly monitor details such as the supply – and return – water temperatures, humidification, water flow, and failed compressors; then remotely make adjustments that meet either the manufacturer's specs – or the user's environment... [increasing] energy efficiency ..., translating to bottom-line savings, and aids in preventive maintenance – trouble can be spotted and corrected off-hours.' (Bowman 1998)

Additionally, technology provides the opportunity for data to be transferred from the sensing devices to a remote monitoring and/or control site. The growth of global communications enables an air-conditioning plant in a Birmingham, UK, office or supermarket to be controlled from Birmingham, Alabama, or from Berlin or Mumbai (Bombay). The world is developing fast in this field. For instance, a development in Calcutta 'is to include two 22-storey "intelli-centres" and 1,200 "intelli-homes", all completely computer integrated and connected to high speed global data networks' (Anon, 1995).

Control

Control, however, let alone 'intelligent' control, needs more than data. It needs to be coupled with experience, another facet of intelligence, which may be human or artificial. Thus when a particular situation pertains (like an increasing pressure in a pipe) and a particular consequence is likely (e.g. damage), an appropriate intervention (e.g. opening a relief valve) may avoid loss or reduction of service. This kind of knowledge-based system is facilitating predictive maintenance.

An article by McCullough (1998) quotes two views on predictive maintenance:

- 'Predictive maintenance solutions that increase systems availability and reduce life-cycle costs, are playing a growing role ...' (Tony Shipley, Entek IRD president and CEO).
- 'Predictive maintenance lets you know what's happening in an asset, so you don't need to actually perform preventive maintenance until it's necessary. It reduces the cost of service because you're pushing out the cost structure ... It's all about maximising the useful life of an asset at an optimum cost' (Tony Newkirk, Director of Service Marketing, Honeywell Home and Building Control).

Learning is something to be done by maintenance and marketing personnel and by the building and its systems, perhaps in what McGregor (1994) called 'the learning building'. Edington (1997) referred to pre-emption – 'spotting the fault before the customer does' – and goes on to quote Chris Conway, the Chief Executive of the Digital Equipment Company: 'Through the application of artificial intelligence we are able, at Digital, to monitor our customers' computer systems remotely and sense up-coming problems. This allows us to anticipate problems before they even happen'.

Complexity and complication

Developments in technology and management have contributed to the complexity of buildings and their care; they have contributed to both problems and solutions. The Intelligent Buildings Conference at the Building Research

Establishment (BRE) in October 1998 demonstrated several instances of solutions for problems not yet identified. The term 'gizmology' conjures such an image. In pursuit of technological wizardry, it is easy to lose sight of the human dimension and to fail to maintain an attitude of care for the relationship between the building and its users. The KISS principle (Keep It Simple, Stupid) has much to commend it. The expression quoted in Chapter 5, 'lots of data; lack of information', is also pertinent here.

As Jim Read (1998), Associate Director of Arup Communications, put it:

> 'A truly intelligent building uses technology to serve, not to dominate. This approach is profoundly different from the early days, when the most "intelligent" building was the one that employed the most innovative technology. "Innovative" soon became "complicated", producing buildings difficult to manage and technology impossible to control.'

The automated building

This was explored in another paper by the author in 1999 (Wood, 1999b). The automated building is many people's perception of an 'intelligent building'. Robert Heller (1990) gave a 'glimpse into the future' description of life in an intelligent building, 'complete with sensory, biometric and personal sensors and scanners'. Many buildings have sophisticated control systems endeavouring to provide relatively static internal environmental conditions and/or to provide security by access control. These may be known as energy management systems (EMS) or building management systems (BMS).

At a relatively simple level, thermostats and timers have provided aspects of an automated building for a long time. Although early systems may not have involved much more than on/off switching, a building can be conceived in which every aspect of its internal environment can be and will be controlled. Implications of control are considered further in this chapter.

Harris & Wigginton (1998) also recognised that:

> 'It is important to acknowledge the difference between intelligence and automation. The latter merely describes an ability to activate and regulate movement, usually with electro-mechanical assistance. It requires instruction from a remote source. True levels of intelligence involve more instinctive responses in relation to local environmental conditions. Therefore, true intelligence relates more to the decision-making process than to the actuation.'

Harrison *et al.* (1998) have identified that in the early 1980s: 'Developers saw the provision of "building intelligence" in buildings and the services made available to tenants as a means of giving their buildings a marketing edge' and noted encouragement given at that time by the Japanese government to 'projects that met its definition of intelligence'.

It is perhaps worth noting the repeated 'needs' for highly sophisticated systems. Sullivan (2001) reflected, however: 'On the downside ... one must confess that many of these products suffer from weak track records. They can be expensive, unreliable and a largely invisible - and thus not very charming - investment'. Further, studies such as DEGW & Teknibank (1992) identified that many buildings deemed intelligent by the technology were not coping with changes in the organisations that occupied them and that such buildings would become 'prematurely obsolete and would either require substantial refurbishment or demolition' (Harrison *et al.*, 1998).

The automated building is predicated on the notion that the environment and occupants are to be controlled; that without intervention, situations will go out of control, that unacceptable conditions will arise. There is a presupposition that the environment is something to be manipulated, even to be fought against, to be brought into line with human acceptability. Indeed, the concept of the building envelope or 'skin' as an insulator of the inside of the building from external conditions - wind, rain, cold - or a moderator between external and internal environments has a long history, as exemplified by Banham (1969) and Burberry (1970). Building Regulations in England and Wales have had requirements for thermal insulation since the 1970s.

Heating and ventilation systems and air conditioning add to the ability to modify the internal environment. Early systems were crude; heating was on or off. Over the years more sophisticated monitoring and control equipment has facilitated a better understanding of what makes for a satisfactory living or working environment and closer control. Thermostats where a mercury switch 'tripped out' to switch on or off hot water service to a radiator automatically when temperature went outside a preset range were revolutionary in their day. These now seem very simple devices by comparison with computer-controlled systems responding to changes and anticipated changes in humidity, occupancy, ventilation, air movement and quality, as well as temperature.

Comfort

Understanding of the conditions that give rise to occupant comfort has also increased. For instance, in the 1970s Fanger (1972) codified the concepts of the Predicted Mean Vote (PMV) and Percentage People Dissatisfied (PPD), subsequently incorporated into ISO 7730: 1994. Humphreys (1976), Haves (1992) and Fordham (2000) are amongst many who have developed these studies further towards defining comfort and determining how it may be achieved. Leaman & Bordass (2000) have identified four clusters of 'killer variables' that lead to a productive workplace:

- personal control
- responsiveness
- building depth
- workgroups.

Leaman & Bordass bring together results from many surveys of buildings and their occupants, carried out by themselves and others, and provide a rich source of material and further reference. It is pertinent to extract a few findings here to give a flavour.

> 'People were more tolerant of conditions the more control opportunities – switches, blinds and opening windows, for instance – were available to them.'
>
> 'In study after study, people say that lack of environmental control is their single most important concern...'
>
> 'In spite of the wealth of research and occupier evidence that high perceptions of personal control bring benefits such as better productivity and improved health, designers, developers, and sometimes even clients seem surprisingly reluctant to act on it.'

Comfort and its relationship to control are discussed later in this chapter.

Automaticity

I am trying to draw out here a difference between an *automatic* and an *automated* response. In the 'automatic' case, an action A always gives rise to reaction B. Thus, for instance, an excessive temperature may always give rise to the opening of a window. This opening could be executed either manually or in an 'automated' way, that is using automation, with some kind of powered or machine-operated device. Key issues with automaticity include:

- when to intervene
- what kind of intervention
- how to make the intervention
- scale of intervention
- who will decide.

The rationale for intervening is also well worth defining; it may give rise to alternative ways of looking at the perceived problem and possible solutions, including 'no intervention needed' (or desired?) or maybe 'not worthwhile'.

Braegar & de Dear (1998) identified that building occupants may respond to uncomfortable or 'unacceptable' situations in three ways:

- acclimatisation
- habituation
- adjustment – personal, cultural or environmental.

Thus, for instance, if occupants are too hot they may respond by:

- over time becoming familiar with the ambient temperature: 'getting used to it'
- coming to accept the 'too hot' condition as satisfactory: 'putting up with it'
- dressing in lighter clothing, taking off a jacket, loosening a tie; dressing and behaving 'as the locals do'; adjusting ventilation, e.g. opening a window.

This means that we should consider a range of possibilities of when and how to intervene. Consideration also needs to be given to the range of building occupants. It may be comparatively straightforward to identify a comfort range for an individual occupant but how to accommodate possibly conflicting needs of a multiplicity of people, perhaps different people at different times? The context in which people are responding may differ. For instance, it is suggested that responses to poor internal environments identified as 'sick building syndrome' (SBS) may be influenced by respondents' general disposition to their work, their employer, home life, etc. (Clements-Croome, 2000, pp. 3–17). 'Work/life balance' is gaining increasing attention.

Smartness

Thus it may be difficult to determine the situation in which an intervention may be appropriate; that is, to define action A referred to above. It is also difficult to determine the appropriate automatic intervention B. Critics of the so-called 'intelligent building' doubt whether it is 'intelligence' that is really being applied. For instance, in an article in *The Economist* (1999) so-called 'smart homes' were described as:

> '... not really all that smart, because what they do is to follow a pre-ordained set of instructions that tells the computer in charge when to switch things on and off. The instructions themselves still have to be programmed to suit individual tastes and desires – and they have to be updated by hand if circumstances change.'

The article identifies a way forward offered by Michael Mozer, a computer scientist at the University of Colorado, '... to get a house to program itself'. Dr Mozer's house has 'smart' systems within it, designed and installed by himself and his students, that use neural networks to recognise patterns of use and work out relationships. Neural networks behave rather like a human brain; they learn from experience, enabling increasingly appropriate responses over time. Thus, when someone comes into a room, 'the system' decides whether or not to turn on a light and how bright it should be. It senses the ambient light level outside and in the room from which the person has entered, it consults its memories related to previous entries, times of day, day of week, etc. looking for, creating and modifying 'patterns of behaviour'. It learns from whether and when the occupant overrides the system's 'decisions'.

The article conveys the flavour of excitement, even passion, with which new technology is sometimes reported. A more recent article (Hawley, 2001) suggests a historical and cultural perspective. He writes from Kyoto, '... a special city, with many ancient wooden buildings intact and in harmony with the newer city ... centuries of zenlike wisdom are infused into the architecture' ... 'That kind of deep value, spanning generations, hardly applies to intelligent technologies yet. There is no zen in my tangled stereo system... Technology at home is not a symphony. It's a cacophony.' Hawley contrasts the architecture and plumbing of Ephesus built in the time of Christ with 'my lousy little bathroom in Cambridge' [Massachusetts]. He compares 'the beautiful Roman latrines' of Ephesus with the nearby 'modern' Turkish town of Kusadasi: 'It's as if more than 2000 years of better living through plumbing had never happened'.

Greenness

As noted in Chapter 5, there is a significant strand of thought that the truly intelligent building is a 'green' building. Such a building is 'one that took the fullest account of passive systems and was "low-tech" – by contrast with the "high-tech" imagery of a highly serviced and automated building with a lot of intelligence to be exercised to keep internal environments "under control"' (Wood, 1999a). This aspect has been developed by the author in further papers (1999b, 2000a, b) and is discussed in more detail in Chapter 8.

It is crucially important to give full attention to matters of building intelligence and care at the earliest opportunity, in the project inception and briefing stages.

The Integer House at Watford was so called to reflect its conceptual basis of being 'intelligent' and 'green'. The building incorporates, for instance, photovoltaic cells on one roof slope and turf on the other. Amongst the features and ideas behind the project (Thompson 1998) are the following.

- The building and its components should be mainstream, not experimental. It should be comfortable and attractive.
- Robust and adaptable.
- Low in resource use, e.g. energy, water, materials.
- Simple central computer control.
- Superinsulated.
- Maximise use of solar energy.
- Rainwater collection, grey water recycling, reed bed filtration.
- Efficient construction, low in waste, high in off-site prefabrication to improve quality and speed.
- Single contact maintenance contract.

Comfort versus control

The work of Haves (1992) has already been referred to; he identified that occupants preferred to be in control of their environments rather than have their heating, lighting and occupation rates and utilisation controlled or monitored remotely. Bunn (1993) has shown that 'people who have greater control over their indoor environment are more tolerant of wider ranges of temperature'. It should be possible to achieve 'ideal' internal environmental conditions through a BEMS and to give individuals control; for instance, to 'override' by opening a window would be to lose control over temperature, humidity, ventilation rates and costs. Evidence suggests, however, that occupants are both responsive and responsible and may be more inclined to turn down or switch off unneeded services than is 'the system'. Work by colleagues at Oxford Brookes University (Humphreys, Nichol, Kessler and McCartney) is ongoing in this area of adaptive comfort.

Leaman & Bordass' four 'killer variables' (2000) – personal control, responsiveness, building depth and workgroups – have been referred to earlier. Further findings relevant here include the following.

- People with window seats tend to be more comfortable (Nichol & Kessler, 1998).
- Occupant satisfaction and productivity tend to decline with increasing building depth; the optimum is around 12 m.
- Deeper buildings also tend to be more complex in terms of building services, uses and user interactions. There is more scope for things to go wrong.
- Room size is a correlate of perceived control for temperature, lighting and ventilation, with perceived control declining with workgroups bigger than about five people (Wilson & Hedge, 1987).
- Occupants have to take account of their colleagues' desires when wanting to make changes; the larger the workgroup, the greater the scope for disagreement and the lower the likelihood of satisfying everyone.
- After three days of putting up with a problem, people either get fed up or give up.
- Shallow plan forms and small rooms with easy, domestic-type controls work best – just like home!

Wyon (2000) argues that delegation of control to users means that they take on responsibilities for which they need insight, information and influence. Insight is about understanding how the building works; information is required in order to learn to use the control, and this must include feedback; only then can the user be given influence. He applies this principle to the development of environmentally responsive workstations, which he terms 'individual microclimate control devices' (IMCDs).

According to Wyon, features of an IMCD must include adjustable local air velocities and heated panels close to the body. The provision of these two features alone allows users to adjust the temperature they experience by as much as 3K above and below actual room temperatures, 'sufficient to bracket 99% of individual neutral temperatures'. This is much higher than normal design expectations of 80% (based on predicted mean vote, PMV) and better by far than that commonly achieved.

Earlier work by Wyon (1993) has demonstrated that productivity improvements from this +/−3K 'degree of freedom' can be as much as 8.6% at the average neutral temperature, and even higher (25.1%) at 3° above that average. This shows the value of providing individual control. Furthermore, as optimal temperature varies for different tasks, e.g. thinking, typing, manual skills, it is even more important to provide means for individual control where a range of tasks is to be carried out in the same environmental space. All that gain will be lost, however, if the ICMD uses a noisy fan, if noise increase is more than 3.9 dBA per K equivalent, 'no overall subjective benefit will be experienced'.

An interesting 'postscript' is provided by Jukes (2000), who was responsible for trying to address a situation where 'a new PC network system ... was planned for forty staff and was failing to cope with eighty staff', even though workload measurement standards showed that 'forty staff should handle the work easily'. In essence, 'sitting someone in front of a PC all day resulted in their personal performance dropping by around 50 per cent without them being aware of it'. The paper describes a sequence of incremental changes introduced to improve the situation, including:

- VDU filters to help tired eyes
- the 'NASA solution' – one large specimen plant per person
- attention to posture
- regular breaks, especially to rest eyes
- air intakes moved to floor to reduce turbulence
- correct humidity (staff were drinking 13 cups of tea/coffee a day and dehydrating)
- ionisers
- polarised light
- antifungal, antibacterial, antimicrobial carpet tiles
- sound-absorbent wall and screen panels to suppress sound
- sound masking to decrease the ambient noise level.

By 'recreating the outdoors, indoors', Jukes and colleagues had produced a 'well building syndrome' at an estimate cost 'equivalent to around 4 per cent'.

The individual versus Big Brother

Robert Heller (1990) identifies that the intelligent buildings that thrive will be those that most closely and completely address the needs and desires of the occupants, which he describes as 'the group ... whose opinion is not often found in print: the Justin Morgans of the world who will ultimately be living and working in the environments created for them'.

A recent study (PROBE, 1996–8), looking at 15 UK buildings that could be considered as 'intelligent', found only one where there was anything approaching 'user satisfaction'. User satisfaction begs questions of who are the users and what are appropriate measures for satisfaction. For instance, a case was reported where for a cleaner to put on the lights for a floor of open offices to be cleaned, a separate number needed to be dialled from each telephone extension to put on the lights for the related workstation.

Gerek (1994) addresses the question of 'How smart should intelligent buildings be?'. For instance, 'Does the owner have a maintenance staff or facilities group capable of operating an intelligent building?'. Who will respond, how and when? Gerek describes two possibilities. For example, 'if normal operating procedure is to call in a service company, then summary alarms at a terminal in the building manager's or information system manager's office may be adequate'. However, 'if the operation is hands-on and the interest is to control events or initiate varying sequences, then individual points may need to be provided...'.

So, who is to be in control? The individual is at one end of the spectrum. Can what he or she wants, or needs, be obtained without conflict with others' wants or needs? To how much control may an individual 'submit'?

Heller (1990) detects movements in attitudes away from concerns about 'Big Brother'. He argues that, although we have been resistant to depersonalisation, our priorities have changed. For instance, we are happier with automatic bank teller machines than with 'standing in line' waiting for a human cashier. Social and political forces have been changing our attitudes. He cites 'the rise in crime and terrorism worldwide', for instance, and of course since Heller wrote in 1990, we have seen great growth in home and telephone banking, insurance and so on. What has this to do with maintenance? It has to do with service; putting the client, the user, at the centre of customer care.

Intelligent service

The traditional approach to building maintenance has been to have a 'handyman' or 'maintenance man' on site or on call. This has changed as buildings and particularly their services installations have become more complex and it has been necessary to either take on a service contract, perhaps with the original supplier/installer, or to call them out. Suppliers and users saw advantage in such relationships and with them warranties or guarantees. Often such contracts have required that equipment be 'maintained in accordance with manufacturer's requirements'. Such 'tie-ins' can be fruitful sources of maintenance work. Alternatively, where trusted relationships do not exist and in an emergency, a tradesperson may be sought using the *Yellow Pages* or similar, with mixed results.

In another sector, food retailing, the Smyth & Wood paper (1995) observed maintenance services being provided by organisations from a range of backgrounds, including security and cleaning companies. Further research (Wood & Smyth, 1996) identified that:

> 'in relation to the JIT maintenance market, the cleaning and security group perceive it as an area that has been easy to enter. The absence of an established product or service line was not an issue ... what was established was a continuous relationship and known track record for cleaning and security services.'

Security companies particularly were able to deploy central computer and telecommunications systems to facilitate rapid response and with recording systems that would give evidence of service levels actually achieved. Furthermore, these companies often had existing links into clients' premises, for instance connecting sensing and monitoring devices and alarms, video recording, etc. to central 'control rooms', often run with military discipline, staffed by accredited, fully trained, attentive personnel.

This communication and control infrastructure provides a significant 'barrier' to competitors to enter the market and gives the potential for competitive advantage for the service provider on both cost and client satisfaction. This may be seen as an integral part of the 'intelligent building' concept.

A similar development has been taking place in the domestic maintenance market. The provision of building maintenance services in the domestic sector from organisations like the AA and Green Flag Insurance has already been referred to. It may be that a growing 'smart homes' market may provide further opportunities. Specialist firms like building contractor McCarthy and Stone, providing sheltered housing schemes, and Tunstall, in the field of personal alarm systems, may be able to build comprehensive care services from their otherwise narrow markets, to serve more fully the increasing elderly, frail, infirm, housebound and home-based populations. A number of projects presented at the 1998 EIBG/BRE conference were targeted at dealing with both 'grey' and 'green' issues.

Market intelligence

Since the 1997 study the AA has revisited its business strategy and has decided to 'return to core business', recognising its key attributes as a motoring organisation. Both the AA and Green Flag had invested, and continued to invest, substantial sums in vetting contractors and it may be that recurrent costs to assure quality were insufficiently supported by the market. It may be that an annual charge of £115 was insufficient to meet the level of service offered.

The majority of AA call-outs were water based, plumbing and central heating related; these areas are susceptible to sensing, through flow monitoring, temperature measurement (especially in relation to freezing), pressure drop and leak detection. Through monitoring and the building up of intervention histories and maintenance profiles, they are capable of informing and being informed by predictive maintenance.

Green Flag demands as a prerequisite of its contracts with householders that central heating systems be subject to an annual service of the boiler at least; it makes sense for that service engineer or his/her company to be one of their 'approved list'. It is possible to see the development of the 'one-stop shop' maintenance service.

Other markets offer other opportunities. For instance, the Private Finance Initiative (PFI) is now bringing forward substantial schemes (hospitals, prisons and education buildings - especially student residences - being in the vanguard) whereby a contractor takes on responsibility for design, construction and operation of the facility for an agreed period of time. This gives substantial incentive to the contractor to forecast and control, amongst other recurrent costs, the need for maintenance interventions and to match these with clients' needs and expectations. This is fertile ground for research, and application of research findings, with publication following once initial commercial and political sensitivities about costs and liabilities have subsided.

Meaning and application

What does this mean for design, construction and facilities professionals?

- While technological advances have facilitated intelligent buildings, inadequate attention to human aspects has tended to leave much potential unrealised.
- Building owners are looking for service and this is being provided by trustworthy and reliable organisations with a customer focus from outside traditional construction disciplines.
- Long-term relationships are being sought; this is hampered by lack of trust within the construction industry and by paucity of information, for instance in relation to life expectancy of building components and in relation to 'organisational churn'.
- Although maintenance is not 'sexy', the market for building care service is substantial.
- The Public Finance Initiative offers great opportunities to combine first costs and recurrent expenditures together with service quality as an intelligent building care package.

Summary

This chapter has investigated some of the possibilities of caring for a building with a strong technological emphasis. In some ways process and product are interrelated: a building that is high in technology lends itself to management and maintenance that are highly systematised and automated. In the next chapter this is considered against a background of increasing interest in issues of sustainability. Possible implications of lower levels of specification and intervention are examined.

References

Anonymous (1995) Calcutta gets 'smart': computer-integrated 'city' planned for Calcutta. *Industry Week* **244**(13), 55.

Banham, R. (1969) *The Architecture of the Well-tempered Environment*. Architectural Press, London.

Bowman, R. (1998) BAS opportunities (building automation systems). *Buildings* **92** (6), 34.

Braegar, G.S. & de Dear, R.J. (1998) Thermal adaptation in the built environment: a literature review. *Energy in Buildings* **27**, 93–96.

Building Research Establishment (BRE) and European Intelligent Building Group (EIBG) (1998) *Intelligent Buildings: Realising the Benefit*. BRE, Watford.

Bunn, R. (1993) Fanger: face to face. *Building Services* **15** (6), 25–27.

Burberry, P. (1970) *Mitchell's Building Construction: Environment and Services*. Batsford, London.

Clements-Croome, D. (2000) *Creating the Productive Workplace*. E. & F.N. Spon, Oxford.

DEGW (London) & Teknibank (Milan) (1992) *The Intelligent Building in Europe*. British Council for Offices, College of Estate Management, Reading.

Economist (1999) The learning home. *Economist* **351**, 90.

Edington, G. (1997) *Property Management: A Customer-Focused Approach*. Macmillan, London, p.88.

Fanger, P.O. (1972) *Thermal Comfort, Analysis and Applications in Environmental Engineering*. McGraw-Hill, New York.

Fordham, M. (2000) A broad definition of comfort as an aid to meeting commitments on carbon dioxide reduction. In: *Creating the Productive Workplace* (Clements-Croome, D., ed.), E. & F.N. Spon, London, pp. 71–76.

Garmonsway, G.N. (1969) *The Penguin English Dictionary*. Penguin, Harmondsworth.

Gerek, M.H. (1994) How smart should intelligent buildings be? *Electrical Construction and Maintenance* **93** (10), 49.

Harris, J. & Wigginton, M. (1998) *The Intelligent Skin: A Case Study Review*. Proceedings of EIBG/BRE conference: Intelligent Buildings: Realising the Benefits, 6–8 October. Building Research Establishment, Watford.

Harrison, A., Loe, E. & Read, J. (1998) *Intelligent Buildings in South East Asia*. E. & F.N. Spon, London.

Haves, P. (1992) Environmental control on energy efficient buildings. In: *Energy Efficient Building* (Roaf, S. & Hancock, M., eds). Blackwell Science, Oxford, pp.39–59.

Hawley, M. (2001) Edifice complex. *Technology Review* **104** (6), 84.

Heller, R.J. (1990) A day in the life of an Intelligent Building. *Journal of Property Management* **55** (4), 53.

Humphreys, M. (1976) Field studies of thermal comfort compared and applied. *Building Services Engineering* **44**, 5–27.

Jukes, J.H. (2000) Optimising the working environment. In: *Creating the Productive Workplace* (Clements-Croome, D., ed.). E. & F.N. Spon, London, pp. 313–319.

Leaman, A. & Bordass, W. (2000) Productivity in buildings: the 'killer' variables. In: *Creating the Productive Workplace* (Clements-Croome, D., ed.), E. & F.N. Spon, London, pp.167–191.

McCullough, T. (1998) Monitors predict commercial building maintenance. *Business First – Columbus* **14** (31), 34.

McGregor, W. (1994) Designing a 'learning building'. *Facilities* **12** (3), 9–13.

Nichol, J. & Kessler, M. (1998) Perception of comfort in relation to weather and indoor adaptive opportunities. *ASHRAE Transactions* **104**, 1005–1017.

PROBE (1996–8) In *Chartered Institute of Building Services Engineering (CIBSE) Journal*, various dates.

Read, J. (1998) *Today's Intelligent Buildings. What Can They Really Offer?* Proceedings of EIBG/BRE Conference: Intelligent Buildings: Realising the Benefits, 6–8 October. Building Research Establishment, Watford.

Smyth, H.J. & Wood, B.R. (1995) *Just In Time Maintenance.* Proceedings of COBRA '95: RICS Construction and Building Research Conference, Edinburgh. RICS, London.

Sullivan, C. C. (1994) High tech/high touch: balancing artificial intelligence and human intelligence in business facilities. *Buildings* **88** (2), 42–45.

Sullivan, C.C. (2001) No drivers wanted (self sustaining buildings). *Building Design and Construction* **42**(4), 7.

Thompson, N.C. (1998) *INTEGER Intelligent and Green Housing.* Proceedings of EIBG/BRE Conference: Intelligent Buildings, Realising the Benefits. Building Research Establishment, Watford, 6–8 October.

Wilson, S. & Hedge, A. (1987) *The Office Environment Survey.* Building Use Studies, London.

Wood, B.R. (1999a) Intelligent building care. *Facilities* **17** (5/6), 189–194.

Wood, B.R. (1999b) *Sustainable Building Maintenance.* Proceedings of Catalyst '99, University of Western Sydney, pp. 129–140.

Wood, B.R. (2000a) *Sustainability and the Right/Left Building.* Proceedings of the Joint Symposium of Conseil Internationale du Batiment Working Commissions W55 and W65, University of Reading, September.

Wood, B.R. (2000b) *Sustainable Building Care.* Proceedings of Conseil Internationale du Batiment Working Commission W70 Symposium: Providing Solutions to Business Challenges – Moving Towards Integrated Resources Management, Queensland University of Technology, Brisbane, 15–17 November, pp. 415–422.

Wood, B.R. & Smyth, H.J. (1996) *Construction Market Entry and Development: The Case of Just in Time Maintenance.* Proceedings of the 1st National Construction Marketing Conference, Oxford, 4 July, pp. 17–23.

Wyon, D.P. (1993) *Healthy Buildings and Their Impact on Productivity.* Proceedings of the 6th International Conference on Indoor Air Quality pp. 3–13.

Wyon, D.P. (2000) Individual control at each workplace: the means and the potential benefits. In: *Creating the Productive Workplace* (Clements-Croome, D., ed.). E. & F.N. Spon, London, pp. 192–206.

8 Sustainable Building Care

In this chapter a substantially 'green' aspect is added. This is often characterised by a 'low-tech' approach to matters. In relation to buildings this may suggest simple structures, perhaps of hand-made or 'natural' materials, and low in services engineering. With respect to maintenance, it may mean avoiding systems that are likely to 'go wrong' or that require sophisticated corrective interventions. The chapter looks not only at the care of a sustainable building but also at the sustainable care of buildings in general, whether or not the building may be dubbed a 'sustainable building'. It provides a 'bridge' to the next chapter, where alternative design constructs are conceptualised.

Sustainability

Sustainability has become a significant area of study in recent years. Initial concerns centred on the relationship, or dissonance, between matters of ecology and economy. As reported by Steele (1997):

> 'A report entitled ''Limits to Growth'', published by the Club of Rome in 1972, focused on the idea of progress and most particularly on the fact that global industrial activity was increasing exponentially, predicting drastic consequences if such growth were not altered. This report, later considered to be naïve ... succeeded in popularising the axiom of ''zero growth''...'

Impetus was given to concerns about continuing availability of resources, especially fossil fuels. Steele goes on to identify:

> 'the first use of the word ''sustainability'' in connection with the environment in 1980 in a publication produced by the International Union for the Conservation of Nature (UCN) ... entitled ''World Conservation Strategy'', in which sustainability was inextricably linked to development.'

It has become conventional to define sustainability by reference to the World Commission on Environment and Development report *Our Common Future* (also known as the Brundtland Report, after the President of the panel) (WCED, 1989): 'The principle that economic growth can and should be managed so that natural resources be used in such a way that the quality of life of future generations is ensured'. This definition has been seen by many as unsatisfactory and many drawbacks were examined and reported by the UK House of Lords (1995). More recent work (Williams *et al.*, 2000), while making reference to a wide range of sources (e.g. Elkin *et al.*, 1991; Farmer, 1996; Papenek, 1995; Smith *et al.*, 1998), still finds it difficult to identify a comprehensive definition. As the author concluded (Wood, 1999), 'It seems easier to identify a commonality of aspects that comprise or contribute to a more sustainable environment'.

Common elements of definitions included:

- 'greenness'
- use of renewable resources,
- reductions in waste, harmful emissions and environmental impact.

For instance, Elkin *et al.* (1991) suggest that:

> 'the most effective mechanism for reducing many of the environmental impacts associated with building materials is to design for durability. Davidson & MacEwen (1983) argue for long-life, high-quality buildings that require little maintenance and are renewed at long intervals rather than low-quality structures that are replaced more frequently.'

The validity of this approach is developed in Chapters 9 and 10.

The sustainable building

What would be the components or features of a 'sustainable building'? Possibilities include:

- 'lean' design
- low (or zero) operational energy
- embodied energy
- low emissions, e.g. carbon
- use of renewable resources
- local materials
- little or preferably no waste
- low maintenance
- high durability
- design attuned to use of building
- adaptability
- reuseability
- use of appropriate technologies.

Vale & Vale (2000, p. 241) reproduce Graeme Robertson's 'required changes to produce a sustainable future' (Robertson, 1993) which generates a 'new paradigm' that includes, for instance:

- understand the limitations of 'design'
- respect material/spiritual connections
- humanising
- design balancing all sensory needs
- objects and systems of long-term value
- holistic considerations throughout
- healthy caring environments
- human-scale design solutions
- male/female design approaches
- unpredictable and perhaps surprising
- respect for the notion of 'craft'.

Although a number of these aspects may be better considered at the design or construction stage of a building rather than once the building is in use, the maximum value of 'building care' is obtained by consideration at all stages. Sustainability needs to be considered throughout the stages of development of the building – in design, in construction and in use. This continuum and the evaluation of costs over time through 'cradle-to-grave' analysis and lifecycle costing are discussed in more detail in Chapter 10. Aspects appropriate for consideration at the design stage of a new building will also be relevant when considering existing buildings for purchase and/or adaptation.

A holistic approach is required and some method for evaluating and adding together or 'trading off' the various aspects that contribute to sustainability. A number of such methods have been developed and data assembled to enable such assessments, for instance BREEAM in UK and NatHERS in Australia. A useful examination of design tools can be found in Pullen & Perkins (1999) and a broader range of issues that might be considered has been developed in the work of Vale & Vale (2000).

Sustainability at the design stage

A number of the issues 'bulleted' above are particularly relevant here and it is at the design stage that many decisions are made that will affect or maybe determine in large part the degree of sustainability of the building's construction and use. Perhaps in the past more attention may have been paid to matters of aesthetics or appearance than how the building may be built, occupied and maintained.

'Lean' design is about 'lightness' or economy (in a broad sense, beyond mere cheapness) of use of materials and appropriate methods and standards. For instance, an external wall would comprise materials that were either renewable or non-destructive of the natural environment or recycled, non-hazardous and non-toxic, not needing unnecessary cutting and waste and with low total energy use, embodied and operational. A 'lean' design would consider the appropriateness and acceptability of a naturally ventilated office rather than specifying air conditioning, which would be expensive in material, installation

and operation. For more detailed study of the subject, consult the work of Graham Treloar at Deakin University, Geelong, and Geoff Outhred and others at RMIT and CSIRO in Australia. There is a Lean Construction Institute hosted by Glen Ballard and Greg Howell of the Universities of Berkeley and Stanford in the USA, at www.leanconstruction.org.

Zero operational energy is now achievable and a practical possibility. Sue Roaf's 'eco-house' in Oxford, built in the 1990s, utilises passive solar gains through a two-storey sunspace and photovoltaic cells to capture and store energy, exporting electricity to the National Grid at times of surplus. Robert and Brenda Vale's 'autonomous house' at Southwell in Nottinghamshire uses thermal mass with thick insulation to achieve a very low-energy solution. Subsequent developments at Hockerton, also in Nottinghamshire, using an earth-sheltered approach, and the Peabody Trust's BEDZED (Beddington Zero Energy Development) in Carshalton, South London, have continued the quest. The BEDZED scheme, designed by Bill Dunster, incorporates living and working environments to also reduce energy used in travel.

Embodied energy is the energy used to make and incorporate materials into a building before it starts to be operational. Interest in this area grew as thermal insulation standards increased. It was believed at one stage that there may be more energy expended in creating and installing some kinds of insulation product, for instance fibreglass quilt, than would be saved over the many years of its 'working' in the building. Much work has been done over recent years to quantify the embodied energy of building products and processes though there is as yet no consensus on what should be included or excluded from the calculations. For instance, a brick will embody much energy that was used in firing it in the kiln. There will also be an amount of energy used in transportation of the brick to site. But what about the energy used in constructing the kiln, extracting the clay from the ground and restoring the clay pit after exhaustion, the energy embodied in building the lorry, the roads on which it runs, the energy expended in incorporating the brick into brickwork, etc., etc.?

Spending time in researching and considering alternatives at the design stage has a great impact on achieving a sustainable building and data is increasingly available to assist in those evaluations. By comparison, little research has been done in relation to the contribution of the construction phase.

Sustainable construction

What facets may be relevant at the construction stage as the project moves from the idea and the drawing board (or its computer equivalent) to the realised building, ready for occupation? Much will depend on the procurement route selected as this confers differing roles and responsibilities on the various participants and affects the scope of their inputs to the process.

For instance, with a 'traditional' procurement process (in the UK the JCT '63/80/98 route), and particularly with new build, architects' inputs will have been most significant at the conceptual and scheme design stages and in preparing the detailed drawings and specifications. From these the building contractor, normally selected by competitive tendering, will be expected to be able to build, with little or no further involvement of the architect. A 'prescriptive' approach to specification, where the architect has given detailed instructions on materials and suppliers, reduces the scope for the contractor to shop around for the best, or cheapest, deal. Thus there has been a trend over recent years to more performance specification, where the contractor is responsible for providing the solution and is able to select more widely. In theory, this could offer increased scope for the contractor to provide from sustainable sources. In practice, pressures of time and profit militate against this and the scope for contractor influence on sustainability may be limited mainly to selection of construction processes. In relation to refurbishment projects, architects are less likely to be involved and surveyors, if engaged, and their clients - often building owners rather than occupiers - are on the whole less interested in sustainability, so there is unlikely to be much pressure on refurbishment contractors with regard to sustainable construction.

With a design/build procurement there is scope for contractors to be involved earlier which increases their ability to influence design decisions in a way that may improve buildability and thus reduce resources required to construct the building. However, anecdotal evidence, particularly from architects, suggests that this approach has tended to 'dumb down' designs, squeezing out unusual or complex features in favour of conventional and simple details that have been tried and tested. This is sensible and safe. However, as long as 'sustainability' feels like something new and risky, it will be difficult to gain the full advantage of working closely together to achieve common aims. Pressures to get to completion as soon as possible, and to reduce build cost, also squeeze out the time to research sustainable construction. Similar considerations apply to contractual arrangements where the building constructor is responsible for the future operation of the building.

Focusing therefore on the scope for adding to sustainability by selection of appropriate building processes, what may a building contractor be able to contribute? Although the range of construction methods available for a building conversion or refurbishment project may be more limited than for a new building, the considerations are broadly similar. In some cases these choices may be more critical to the overall achievement of sustainability; after all, the very reuse of an existing building is of itself contributing to sustainability. Conventionally contractors' inputs have been categorised under the general headings of plant, labour, materials and finance.

Plant

- Restricted access for large items of plant
- Noise, dust, nuisance within occupied areas
- Energy used in operation

- Resources consumed in manufacture of the equipment
- Life expectancy of the plant
- Downtime, being loss of efficiency of resource use

Labour

- General tendency, in the Western world, to reduce labour needs may result in greater use of more, and more expensive, and resource-intensive plant.
- However, refurbishment projects often feature more complex decorative features, especially in heritage properties, and these are often more labour intensive.
- In some parts of the developing world, for instance in Africa, there is renewed interest in labour-intensive construction, valuing the individual and the 'dignity' of work (McCutcheon, 1999; Taylor-Parkins & McCutcheon, 1999).
- The local availability and development of an appropriately skilled workforce is also pertinent.

Materials

A number of issues such as renewability and recycling have already been discussed above. Further considerations may include:

- availability of local supply
- reduction in cutting and waste
- susceptibility to damage
- fixings to facilitate reuseability
- durability and maintenance needs
- decoration and its possible avoidance.

Finance

Building projects, including refurbishment projects, are to a large extent financed by building contractors, at least for the duration of the construction work. Although it will be common for there to be stage payments through the contract, the contractor has to finance the work 'up front'; cashflow is as important to financial success as ultimate profits. Financial requirements are closely related to procurement method and the relationship between procurement and sustainability has been discussed elsewhere. Suffice it to say here that the greater the exposure to risk as the contractor and financiers perceive it, the greater the cost of finance. The more prescribed the materials and methods of construction, the more constrained the contractor's room for manoeuvre and therefore the more expensive the project. Thus 'sustainable construction' is likely to be more expensive than 'normal' until such time as it becomes itself the norm.

Sustainable care

By comparison with the design and construction phases, scope for 'adding' or 'converting to' sustainability once the building is finished is much more constrained. Earlier decisions, for example choices of built form, materials and assembly details, will be important influences upon what maintenance will be required, how often and by whom. For instance, a design decision to carpet a staircase and a construction decision to stick the carpet down with adhesive will have ramifications both in terms of cleaning regime and also time and expense involved when the treads become worn and require replacement. A decision to install a particular fan may require it to be serviced at specified intervals by an approved engineer.

Sustainable care is about more than the care of a sustainable building; it is also about caring for any or all buildings in a sustainable manner. Determinants of sustainable care could include:

- low maintenance
- high durability
- design attuned to use of building
- adaptability
- reuseability
- use of appropriate technologies.

For some time, people have claimed products such as PVCu windows and rainwater goods as 'maintenance free' and postulated the possibility of a maintenance-free building. Although not completely free of maintenance (requiring attention to ironmongery and seals, and cleaning out of leaves), the self-finish nature of the materials is such that they are not demanding of decoration; thus costs may be saved both at installation and through the life of the building. How 'sustainable', though, are these low maintenance products? Calculations show PVCu to have much higher embodied energy than timber; aluminium is also very high. PVCu is an oil-based product, a non-renewable resource; aluminium is a product of bauxite, which requires very high temperatures and heavy industrial plant to convert. Timber, on the other hand, is a naturally regenerating product and available close at hand in most parts of the world, but prone to variability and to rot. Appropriate selection of species, and pieces without shakes, knots and other defects, together with detailing to avoid excessive exposure, will assist in reducing movement and dimensional change – warping, shrinkage, swelling. Preservative treatments, paints and stains also have much to offer in lengthening redecoration, repair and replacement cycles, although serious doubts have been raised in relation to their toxicity and other issues related to 'environmental friendliness'. Low maintenance and sustainability are by no means necessarily related; indeed, they may be in conflict.

Durability

Perhaps the use of *high-durability* products will enhance sustainability? Alex Gordon's (1974) concept of the long life/low energy/loose fit building is relevant here. The concept is described more fully in Chapter 9 but it is sufficient to say here that partly in response to the oil crisis of that time and increasing awareness of environmental issues, Gordon was expounding an approach intended to be light on use of resources. Others have taken up this theme. Davidson & MacEwen (1983) incorporated 'long-life buildings' in their recipe for the 'liveable city' and for Elkin *et al.* (1991), durability was 'the most effective mechanism' for reducing environmental impacts.

Leaving aside for the moment questions about how durability is to be assessed, it can be seen that a 'durable' product or material will not necessarily be 'sustainable'. Let us take the window again as an example. An aluminium-framed window may be expected to have a long life, possibly in excess of 60 years, as may a steel window provided that its galvanising coat is complete and is kept protected by adequate paintwork. However, many aluminium windows of the 1960s and 1970s failed because their sections and joints were insufficiently robust, their 'mill' finish became pitted by the urban atmosphere and spiral balance mechanisms seized up. Steel windows are vulnerable to rust at fixings where galvanising is incomplete and easily damaged and are not easily 'adjusted' when the frame becomes distorted; they are not readily recyclable. In contrast, windows of the readily renewable resource, timber, are of very variable durability. The life expectancy of a wooden window will be determined by a multiplicity of factors, including:

- species of timber
- sectional details
- joints
- exposure
- maintenance
- decoration.

It is a common assumption that a hardwood window will be more durable than one of softwood. However, some hardwood species, such as box and poplar, are less dense than softwoods; the term 'hardwood' only denotes that the timber is from a deciduous, broad-leaved tree. There will be issues related to the source of the timber, whether it is from a 'sustainably managed resource', and this can be difficult to substantiate. There have been attempts at creating schemes, with varying success, for effective labelling of timber from managed forests. Impetus here stems from legitimate concerns about 'cutting down the rainforest' but this has resulted in a retreat from use of tropical hardwoods with consequential losses to economies based on the growing and trading of plantation timber. The rekindling of interest in understanding the characteristics of locally grown timber, however, is a welcome development.

The performance of a window frame is influenced largely by its dimensional characteristics. For instance, too slim a cross-section will result in a greater tendency to twist; too shallow a slope on the weathering faces or lack of drip-mould will leave a cill with more precipitation left on it and in it. Poor decorative state will enhance problems. Timber will decay from constant wetness or through repetitive wet–dry cycles and this will also affect joints that are poorly designed or made. In wet climes it is particularly important in the quest for durability therefore to protect timber windows as far as possible by deep overhanging eaves and by using substantial and well-detailed cills. Whether that increased amount of material is then 'saved' by greater life of the window could make for an interesting piece of research.

In relation to exposure of the window, it may be that the building design could incorporate more windows on less exposed elevations. Elements at a height are also generally more exposed than those at lower levels, so again attention to building form can help promote greater durability. Windows on fewer facades and a lower building may also facilitate easier and less resource-intensive cleaning of windows. Floor/wall and window/wall ratios are recognised as critical measures of relative construction costs of different design options, so it may be that similar simple calculations could be developed as 'rule of thumb' measures of susceptibility to sustainable care.

Use

Not only will dimensional aspects of the building and the quantity and disposition of different elements and materials have important implications for its sustainable care, but so too will the type and intensity of use of the building. This will affect many aspects of the building, but most obviously perhaps the wearing surfaces.

- Hinges and opening and closing mechanisms to doors and windows in intensively used buildings will come under stress; they need to be selected with this in mind. Alternatively, it may be that some of this continual opening and closing could be relieved by installation of electromagnetic 'hold open' mechanisms, for instance in a place of assembly. These will hold doors open while the building is in use, while closing automatically in the event of a fire being detected. This will seriously reduce the number of hinge movements, thereby reducing wear and potentially prolonging life of the component. Whether this is a net gain in terms of sustainability could be a complex calculation.
- High intensity of use, and use by large numbers of visitors who do not 'own' or care for the building, may result in higher levels of damage, for instance knocked and damaged walls, especially at corners in corridors and other circulation areas. Whether it is more sustainable to 'up' the specification, to include buffer strips at dado level and corner protection strips, especially if

these are in plastic, or to patch plaster and redecorate may depend on the frequency of such action. There may also be a 'no action' option, although not to repair damage may induce a feeling that no-one cares, with the possibility of further damage and a spiral of decline.

- Floors will wear more or less in proportion to the amount of traffic; thus thicker vinyl, linoleum or thermoplastic tiles or sheet may be warranted although these materials may not score well on sustainability. Many adhesives are high in unhealthy 'volatiles'. It may be appropriate to make more use of self-finished screeds, if aesthetic, acoustic and other objections can be overcome, or timber boards if a smooth, continuous finish is not essential. Cleaning regimes may be important factors, for instance for a commercial kitchen. Many cleaning materials are also quite dangerous to users, requiring care in their application and in storage.
- By contrast, sanitary fittings, perhaps because they are designed to deal with dirt and waste, seem to be highly durable and to offer long life. They do not generally give a lot of problems in use. Often sanitary appliances will be replaced not because they are broken, although chipped enamel or hairline cracks may be unsightly, but because their style or colour has become unfashionable. There may occasionally be breakages, most likely due to vandalism. Blockages, although hopefully infrequent, also cannot be endured; they are unpleasant and must be dealt with promptly. What 'sustainable care' has to say to sanitary installations is probably focused mostly on the sustainable use of water. Low-flush cisterns, 'grey water' systems that make use of recycled waste water, showers instead of baths and use of dry composting toilets may make for more sustainable systems providing that their ongoing maintenance requirements are not greater than those that they replace.

A number of aspects of durability, as discussed above, relate to situations where intensity of use suggests the specification of a highly durable product. However, there will also be situations in which the deployment of a component of high durability may be regarded as 'overspecification'. It is contended that what is appropriate is design and specification attuned to the anticipated use of the building. Therein lies a problem: how to anticipate the type and intensity of use? Often decisions will be based on experience, which may not be extensive. Design details may be derived from those developed for a previous, similar scheme, perhaps modified in the light of subsequent experience. Sometimes, extensive research may be appropriate to design effectively for a new or unfamiliar use.

Radical review and change

Sometimes, the way a service is delivered within a building comes under radical review, resulting in a need for a substantially different relationship between the building and how it is used. For instance, the Victorians built large sanatoria for the incarceration and treatment of the mentally insane, out of sight in rural settings. More recently, a more humane approach, with the assimilation of people with mental problems locally through 'care in the community', has promoted use of more domestic environments and rendered the Victorian buildings redundant. It has been difficult to find new uses for the old buildings but the sites are often very attractive for the development of new housing estates. The redevelopment of the former Littlemore Hospital by Berkeley Homes to create St George's Park on the edge of Oxford has already been referred to.

On the whole, larger spaces lend themselves to more opportunity to adapt; uses can grow, reduce and change within the space. Lightweight partitions can be readily constructed and deconstructed to divide and re-divide the larger space. By contrast, to create one large space from two or more small, purpose-built rooms can be very disruptive. If load-bearing walls are to be removed, new joists or lintels will need to be inserted to support the structure above and much 'making good' will be required. Framed structures provide benefits here; they allow flexibility in the organisation of floorspace between and around their columns. The downward projection of beams into spaces is not generally problematic other than where it becomes necessary to install new services. Forethought at the design stage and the provision of service zones above suspended ceilings will normally overcome this. Open-web designs such as 'castellated' beams, 'metsec' joists or trussed girders allow for the passage of pipes and wires. The vertical passage of services from one floor to another can be more problematical and for this dedicated service zones, dry risers, etc. can offer useful opportunities for accommodating new and additional services. These can be provided in association with 'service cores' of kitchens, toilets, etc. where water and waste services are concentrated. Such cores are a good way of using excessive depth in a building. Natural light falls off with increasing distance from windows and occupants tend to prefer and be more comfortable with window seats (Nichol & Kessler, 1998; Leaman & Bordass, 2000). Thus though large spaces may be in theory more adaptable, there tends to be a practical limit on building depth, which Bordass (1992) estimated to be about 12 m.

Adaptability

Thus the depth and arrangement of rooms and supporting structure may be critical determinants of the adaptability of a building to changing needs over time. Other factors will also impact upon the potential for sustainable care of the building, if it is to be capable of changing use and reuse, either as a whole or in its component parts. Buildings generally will not be infinitely adaptable; whilst it may be possible in theory to construct such a model, cost is likely to be a dampening factor, especially against uncertain take-up. Some buildings may be incapable of conversion to accommodate a new use due to, for instance, planning constraints or structural incapacity. Buildings as they age tend to decay, especially the external, weather-resisting envelope. There may also be structural cracking and subsidence due to movement. These will be inhibiting factors in deciding whether an existing building is worthy of reuse.

On the plus side, a building may have architectural merit and may over time accumulate cultural or heritage value. The presence of architectural features may encourage retention of an old building for reuse while inhibiting how, in detail, it may be reused. There may be situations in which, even if the whole building cannot be reused in its present arrangement, some parts can be retained and some components recycled. At the 'heritage' end of the spectrum, there is a good market in architectural salvage of period features and at the more mundane end, there is growing interest in recycled material, doors, frames, skirtings, joists, as people become more interested in sustainability and land-fill tax starts to bite. If not sold or reused immediately, components could be put into storage for subsequent reuse. It is postulated (Thompson *et al.*, 1998) that there may be limited opportunity for the reuse of standard components, for instance in UK hospitals, where the National Health Service has for a long time encouraged development of standard 'parts' and systems. It will be important that the recovery of elements for possible reuse does not of itself require additional energy and resources, including for instance transporting to and from storage and creation of additional built space for that storage.

Use of technologies appropriate to the building and its users and the skills of those responsible for its construction and care will also be important. For example, although it may be possible to demonstrate a high sustainability quotient for a design based on prefabricated concrete panels, this may not be conducive to sustainable care. There will need to be a continuing technical resource available to diagnose and rectify defects and to undertake structural alterations into the future. The term 'appropriate technology' is often used as a more acceptable alternative to 'low technology', particularly in relation to developing countries. The intent is to avoid the suggestion that 'low' is somehow inferior. In that 'high technology' may be the best solution in a particular circumstance, the term 'appropriate' is appropriate!

Recognising, however, that 'high technology' requires skills in design, development, implementation and use that are implicitly new and non-traditional, these skills are likely to be relatively scarce. Training, teaching and learning are likely to be needed. By contrast, 'low technology' implies the use of materials and related skills that are more readily available in more people. Thus low technology makes sustainable care more accessible. Training may still be required, though this is likely to be less demanding. There is also a case to be made for low technology as a device for 'empowerment' of people, important for development at a personal level, and also supportive of individual and corporate 'ownership' of a building and its loving care.

The autonomous building

One approach to limiting resource utilisation in the creation and care of a building may be to expect it to in essence live off its own site and the resources therein and thereon: an 'autonomous' building. By contrast with the 'automated building' considered in Chapter 7, the autonomous building may be considered to be a 'greener' and therefore more acceptable concept although that suggests a number of questions, some of which are addressed here.

A definition of building autonomy is provided by Robert and Brenda Vale (2000) in relation to their own autonomous house: 'The autonomous house on its site is defined as a house operating independently of any inputs except those of its immediate environment. The house is not linked to the mains services of gas, water, electricity or drainage...' (p. 7)

As suggested in a previous paper (Wood, 1999), the notion of a building that takes little or nothing from the environment and is low on building services seems to be very much in line with the concept of sustainable building. The 'lightness on the earth' represented by the Vales' autonomous house may seem more readily achievable for single houses or communities in rural and extensive environments and caricatured as being designed and populated by 'ageing hippies'.

An aspect of autonomy worthy of further consideration is the determination of the line or boundary within which autonomy is sought or measured. The idea is realistic and realisable that a building should be:

- constructed purely of materials from 'on site'
- use power derived only from the sun, wind and water that falls on the site, with
- the occupants growing their own food, making their own clothes
- dealing with their own waste, and
- dependent on nothing and no-one from outside.

Once interaction with a wider community is required, the issue becomes more complicated than 'self-sufficiency' suggests and is beyond the scope of this book.

The principle of the autonomous house as espoused by the Vales above is worthy of further deliberation. Detailed theoretical analysis of the technical options was carried out. Thus, for instance, in relation to heating (pp. 94–5), the emission of an adult male engaged in various activities is deduced from CIBSE data, ranging from 115 watts seated to 560 watts for 'pick and shovel work'. The combined sensible and latent heat output of a 3.0 kg cat is presented as 14.8 watts. The Vales calculate, after correcting for size and weight of typical cats, that at a 'packing density' of 15 cats per square metre, a room of 17.5 m^2 would contain 260 cats 'with a heat output of nearly 4.0 kW', a 'useful heat gain'. The Director of the Office of Energy Conservation in Boulder, Colorado, is quoted to report: '... Canada's two-storey Saskatchewan Conservation House in Regina is so well insulated that according to its designer, it needs "only a single light bulb and two couples making love" to heat it'.

In Australia, Michael Mobbs (1998) has demonstrated the achievement of a 'sustainable house' on a 'step-by-step' basis, renovating an existing late 19th-century house in Sydney and making the household 'self-sufficient in water, waste and energy systems'.

Perhaps there are principles and practices related to the autonomous house that could be extended to a wider range of building types. For instance, Vancouver City apparently subsidises special easy-to-install barrels for rainwater collection from downpipes for use on lawns, gardens and patio plants. Designed for use with hose or cans and with child-proof openings for safety, they expect to save around 40% of total household use – 1300 gallons in peak consumption summer months. Careful attention to detail, giving quality time to research at the conceptual design stage, could potentially lead to an independent or 'care-free' building, or at least one low on care needs.

Are some building types and users intrinsically more likely to be more demanding of maintenance than others? Increased or heightened demand may be manifested in terms of more frequent maintenance, or attention by more highly skilled personnel, who may also be in short supply. Inattention, inadequate or improper attention may lead to substantial and expensive 'downtime' and consequential losses in the event of failure of a particular service or facility. In the case of the automated building, the effects of 'failure' may be catastrophic and it is common to have available back-up systems, emergency power supplies, etc. By contrast, the autonomous building should be functioning within the constraints of lower demand and a higher toleration of variation.

The autonomous house has been realised. Can an autonomous office be achieved or an autonomous hospital or factory? To an extent this is a matter of scale and perhaps of vision. Energy efficiency has been a good example of how much progress can be made. A review of the real requirements of buildings and of their occupants may reveal scope for worthwhile reductions in running costs. Although 're-engineering' has a violent rhetoric that sits somewhat uncomfortably with more sensitive images of sustainability, it has much to commend its underlying theme of going back to asking the basic questions and challenging assumptions. For instance, do we need offices as we have come to know them? The 'home office' or 'tele-cottage' is already with us. However, it is suggested that 'the office' is still required, to provide for the social function previously provided perhaps by the pub, club or church – somewhere to meet friends, share problems, plan and discuss events. In these situations both home office and office as social centre have become more similar to each other, lending themselves to a kind of 'homely' residential type and scale of construction, with brick, timber and tile more likely than steel and glass.

Similar trends can be seen with other building types. For instance, in the UK health service, there have been trends towards less 'intrusive' treatment and shorter stays in hospital. It has been possible to reduce the number of beds and to focus more on domiciliary care. More domestic scaled environments, 'homely' in appearance, are particularly favoured for treatment of mental health and by the private sector.

In respect of supermarkets, often in the UK constructed with brick-and-tile exteriors to look 'domestic', technology together with the bulk purchasing that spawned their growth is now enabling a return to smaller-scale outlets in village communities and the ability to pick up one's tele-order without the need for extensive sales floors. It may be that technology is now facilitating simpler and more sustainable buildings, and simpler maintenance.

Users and user needs

In the case of a single dwelling, it may seem obvious that the needs of the occupant(s) would be a paramount consideration. But even then, how much should future needs be considered, especially if they conflict with present needs? What about the possible needs of possible subsequent occupants? What if owner and occupant needs or perceptions differ? In the case of larger buildings and with a multiplicity of owners, lessees and tenants and individual users, there may be a very wide range of needs and opinions. It may be tempting to argue that therefore it is not worthwhile to identify needs or canvass opinion and there is certainly strength in the argument that unless action can be taken, it is better not to ask.

At the BRE Intelligent Building Conference already referred to in Chapter 7, an instance was reported where, in order to clean a floor of offices, the cleaner needed to dial up from each telephone extension to put on the lights for the related workstation. Perhaps the cleaners would not have been considered as 'users' of the building and therefore their views were not canvassed at the designing or fitting-out stage. However, such experience can be elicited through a post-occupancy evaluation (POE) exercise and fed in through a feedback loop to future designs and to reprogramming (or, *in extremis*, rewiring) of the building under review.

Alternative levels of care: chronic and acute

A shortcoming of the sustainability debate is the focus on the achievement of the 'sustainable building' or of 'sustainable development', suggesting that a proposed development is or is not sustainable – an absolute measure. Perhaps there is a threshold, a point beyond which a proposal tips from being sustainable to not being. Perhaps some criteria may be satisfied and others not. Perhaps there could be a kind of 'conditional sustainability', a situation where sustainability may be achieved provided that certain interventions are made at some time(s) in the future.

A building could be conceptualised whose construction required such inputs of 'renewable' materials, labour or energy that even in a lifetime of low or zero further inputs, the original 'investment' could not be considered to be 'sustainable'. At the other end of the spectrum, a building so consuming of resources during its operation that it could not be considered sustainable is very easy to conceptualise; perhaps this is the 'norm' of modern construction. In between the extremes there are degrees of investment that could be considered to relate to a 'sustainable building'; that is, one that can be created and operated sustainably. The next paragraphs explore two 'care' paradigms borrowed from nursing: *chronic* and *acute.*

> 'Chronic': adj. 1. (of a disease) affecting a person for a long time. 2. Having had an illness or a habit for a long time, a *chronic* invalid (Hawkins, 1979).

For building care, this could be considered to be an enduring situation, for which either palliative care or even 'no action required' may be an appropriate response. It could be that unless an intervention to 'improve' or alleviate the condition is warranted or demanded, this state is 'livable with' and, as the definitions above suggest, for a long time. It could be that, from the building's 'completion' (of construction) and first occupation or from a subsequent point, there is a condition of 'managed decline'. This is a valid approach to 'care'.

> 'Acute': adj. 1. Very perceptive, having a sharp mind. 2. Sharp or very severe in its effect, *acute pain, an acute shortage.* 3. (of an illness) coming sharply to a crisis of severity, *acute appendicitis* (Hawkins, 1979).

In building care terms, this is a situation demanding urgent attention. One response would be to effect a temporary 'running repair' to restore service, albeit perhaps at a lower level of performance. Getting beyond the crisis point is the immediate priority. There is a possibility of anticipating such crises, either as far as knowing that such-and-such a 'failure' will occur, perhaps periodically, but uncertain as to timing, or with modern technology within predictable limits. ('Just-in-time maintenance' is relevant here, as discussed in Chapter 4.) In such circumstances it would be possible to instigate an intervention designed to maintain or even improve upon the previous service level.

From a perspective of sustainable care, both chronic and acute regimes are valid. What becomes possible with a 'sustainable building' is a strategy of minimal intervention where as little as is necessary is done as late as possible, thus reducing the resource input to the lowest level commensurate with maintaining a standard of performance satisfactory to building users.

Autonomous building care

The autonomous house as envisaged by Robert and Brenda Vale, with its own solar panels, wind turbine, rainwater filters and composting toilet, is not without maintenance needs so how are they to be met? Some of the questions appropriate to ask in relation to the automated building are equally applicable here. How sophisticated are the systems? For how long will parts be available?

Can the systems be maintained easily by unskilled personnel, perhaps the building's owner or occupier?

Autonomous maintenance of a building suggests a need for on-site personnel to execute necessary maintenance; this in turn suggests either suitably skilled personnel or equipment appropriately selected to meet the capabilities of site 'staff', implying perhaps low-maintenance equipment. Such equipment would also need to have recognisable failure patterns and clear instructions on remedial or restorative actions to be taken in response to such events. The latter, low-maintenance solution is perhaps a more sustainable option than the previous, predicated on personnel with certain skills, for when personnel leave, as surely they will, similarly skilled replacements will be required and they may be difficult to source at the time.

Approaching sustainable building care

A sustainable building and sustainable maintenance are possible; indeed, the one presupposes and requires the other. Perhaps the key elements here are a general 'lightness' or economy in the use of materials, use of renewable resources and a congruence of anticipated life of the building with that of its component parts.

While zero maintenance may be striven for, it is ultimately unachievable and what is desired is a situation where all will finally fail simultaneously. Statistically that too is unlikely and *some* maintenance will be either necessary or desirable to reach a point at which the building's 'death' is acceptable, the occupant(s) being ready to move on to a new building.

Alternatively, a 'minimum total maintenance' model may be pursued, in which a certain level of maintenance is carried out that will sustain the building in suitable condition for a period that will justify that maintenance. It is also possible to determine a further point in the continuum of maintenance activity where the 'life' of the building may be maximised. This maximum maintained life may be attained by a combination of repairs, cleaning, fine tunings, renewals and upgradings to enable the building to meet the demands made of it.

In some situations flexibility in how the building may be used is an important facet in its sustainability. In other instances that flexibility may be provided in such a way as to demand a high level of maintenance.

A number of sustainable building models can be constructed which imply different maintenance regimes. These regimes may be themselves sustainable, but not necessarily. Those described in this chapter have generally featured low maintenance. There will be circumstances, however, where a relatively high level of maintenance may be more appropriate, for instance where very high levels of cleanliness may be critical, such as a hospital or a commercial kitchen. However, it may be that existing levels of maintenance expenditure on buildings as a whole may be unsustainable.

The Barbour Report (1998) estimated that the UK building maintenance market was worth £28 billion, compared with £10 billion for new build. Can or should building maintenance be sustained at this level? At the same time, more than 1.5 million dwellings were recorded in the English House Condition Survey as being 'unfit' (DETR, 1997). Should maintenance be increased?

Alternatively, should new-build expenditure be increased, hopefully facilitating lower maintenance expenditures in the longer run? There is a strong argument that at present much expenditure is being directed towards maintaining buildings that are beyond economic repair and that insufficient attention is being given to achieving sustainability.

Value and values: today and tomorrow

> 'From now the pound abroad is worth 14% or so less in terms of other currencies. It does not mean, of course, that the pound here in Britain, in your pocket or purse or in your bank, has been devalued.' (Harold Wilson, 19 November, 1967)

Of course, with inflation, a pound today buys less than it did last year, let alone in 1967, but that is rather an oversimplification. For instance, whilst an asset's value may be depreciated over time, scarcity is such that with property, for instance, its value may appreciate over time. Economists have developed concepts and mechanisms such as Net Present Value, Internal Rate of Return and Discounted Cash Flow, and these and others are covered in textbooks aimed at students of building finance and quantity surveying; they are not discussed here. What is important here is that the pound, dollar or euro of today will be worth something else in the future. Values change, and not just monetary values.

Building developers and their agents and advisors are either explicitly or implicitly making judgements upon where to spend today's and tomorrow's monetary and other resources. How much is it 'worth' to invest today's money in buying forward a better future? How much is needed to rectify mistakes and results of low investment in the past, perhaps by previous generations? The WCED (Brundtland) definition of sustainability has already been stated at the beginning of this chapter; it is worth reading on.

> 'Sustainable development is not a fixed state of harmony, but rather a process of change in which the exploitation of resources, the direction of investments, the orientation of technological development and institutional change are made consistent with future as well as present needs.' (WCED, 1989).

Note, 'future *as well as* present needs' (my emphasis). The future is not necessarily more important a consideration than the present, nor is it less important; both present and future need to be considered. Not only is the quality of life of future generations to be ensured, but that of the present too. It is hard to find, let alone deploy, resources to correct the errors of the past and meet the needs of today's and future generations. How are the needs and wants of the future to be determined? Forecasting is difficult. 'The future' is surveyed further in the chapters that follow.

Summary

This chapter has discussed a number of aspects of sustainability and their application to buildings. It has explored the maintenance of sustainable buildings and the sustainable maintenance of buildings, whether they be particularly sustainable or not. A number of possible scales of maintenance interaction have been explored and it is postulated that it is not necessarily the lowest maintenance regimes that represent the most sustainable options.

There are choices to be made and the options selected will be influenced strongly by the values attributed and the estimates and assumptions made about how materials and components of buildings may deteriorate over time and what may be the expectations of buildings in the future.

The importance of relating maintenance regimes to the skills and propensities of building users has also been covered. Maintenance will only be sustainable if personnel have the requisite skills and/or if components and systems are installed that are easily maintainable. There is a role for education here; maintainability and sustainability are often accorded low priorities.

The chapter which follows synthesises such matters and introduces two major design archetypes: the LEFT and RIGHT buildings.

References

Barbour Index (1998) *The Building Maintenance and Refurbishment Market: Summary.* Barbour Index, Windsor.

Bordass, W. (1992) Optimising the irrelevant. *Building Services*, **14** (9).

Davidson, J. & MacEwen, A. (1983) *The Liveable City in the Conservation and Development Programme for the UK.* Kogan Page, London.

Department of the Environment, Transport and the Regions (1997) *English House Condition Survey, 1996: A Summary.* DETR, London.

Elkin, T., McLaren, D. & Hillman, M. (1991) *Reviving the City: Towards Sustainable Urban Development.* Friends of the Earth, London.

Farmer, J. (1996) *Green Shift: Towards a Green Sensibility in Architecture.* Architectural Press, Butterworth-Heinemann, Oxford.

Gordon, A. (1974) Architects and resource conservation. *RIBA Journal* **81**(1), 9–12.

Hawkins, J.M. (comp.) (1979) *The Oxford Paperback Dictionary.* Oxford University Press, Oxford.

Leaman, A. & Bordass, W. (2000) Productivity in buildings: the 'killer' variables. In: *Creating the Productive Workplace* (Clements-Croome, D., ed.). E. & F.N. Spon, London, pp. 167–191.

McCutcheon, R.T. (1999) *South African Musclepower: A Great Force in Nature*. Proceedings of Customer Satisfaction: A Focus for Research. Joint Symposium of Conseil Internationale du Batiment Working Commissions W55 and W65, University of Cape Town, pp. 813–822.

Mobbs, M. (1998) *Sustainable House*. University of Otago Press, Dunedin, New Zealand.

Nichol, F. & Kessler, M. (1998) Perception of comfort in relation to weather and indoor adaptive opportunities. *ASHRAE Transactions* **104**, 1005–1017.

Papanek, V. (1995) *The Green Imperative: Ecology and Ethics in Design and Architecture*. Thames & Hudson, London.

Pullen, S. & Perkins, A. (1999) *Design Tools for Assessing the Environmental Impact of Houses*. Proceedings of Catalyst 99, University of Western Sydney, pp.105–116.

Robertson, G. (1993) *New Paradigms for Ecobalance*. Proceedings of New Zealand Planning Institute Conference, University of Auckland, Centre for Continuing Education, p.26.

Smith, M., Whitelegg, J. & Williams, N. (1998) *Greening the Built Environment*. Earthscan Publications, London.

Steele, J. (1997) *Sustainable Architecture*. McGraw-Hill, NewYork.

Taylor-Parkins, F.L.M. & McCutcheon, R.T. (1999) *Choice of Technique Analysis: Decision Making Tools for Employment-Intensive Road Construction*. Proceedings of Customer Satisfaction: a Focus for Research. Joint Symposium of Conseil Internationale du Batiment Working Commissions W55 and W65, University of Cape Town, pp. 794–802.

Thompson, D.S., Kelly, J.R. & Webb, R.S. (1998) *Designing for Short Life: Industry Response to the Proposed Reuse of Building Services Components*. Proceedings of COBRA '98: RICS Construction and Building Research Conference, Oxford, 2–3 September. RICS, London, Vol. 2, pp. 122–132.

UK House of Lords (1995) Report of the Select Committee on Sustainable Development. HMSO, London.

Vale, B. & Vale, R. (2000) *The New Autonomous House*. Thames & Hudson, London.

Williams, K., Burton, E. & Jenks, M. (2000) *Achieving Sustainable Urban Form*. E. & F.N. Spon, London.

Wood, B.R. (1999) *Sustainable Building Maintenance*. Proceedings of the Australian University Building Education Association (AUBEA) 24th Annual Conference and Catalyst '99, University of Western Sydney, Australia, 5–7 July, pp. 129–140.

World Commission on Environment and Development (WCED) (1989) *Our Common Future*. Oxford University Press, Oxford.

9 Building Futures: Left or Right?

Having identified and discussed in the previous chapter a number of facets of sustainability, this chapter proceeds to conceptualise two radically differing approaches to the attainment of a sustainable building: the RIGHT building and the LEFT building. Automation and autonomy or self-sufficiency as influential and driving factors are discussed.

Introduction

This chapter will focus on conceptual futures, contrasting the 'automated' building and the 'sustainable' building. It examines how buildings may be designed for the future and their maintenance needs evaluated. For many, the automated building is synonymous with an intelligent building. For some there is a focus on devices and control and possible remoteness, implying automation with varying degrees of intelligence; for others, the truly intelligent building would be a 'green' building.

Features or attributes that define the automated building may include sensors, automatic or automated responses, responsiveness, intelligence and control. Sensors come in many forms. For instance, in relation to fire there may be smoke detectors, heat detectors or heat-rise detectors and these will be connected to sprinklers and visual or audible alarms in the vicinity and/or elsewhere within the building and perhaps connected directly to the fire brigade.

Determining the appropriate degree and timing of response can be important. For instance, a smoke detector that is over sensitive will be set off when bread is toasted too severely. This is inconvenient, especially if it goes off several times a day and requires evacuation each time. However, the situation will be much worse if occupants tamper with the detector or ignore alarms, with potentially disastrous consequences.

There are also matters of individual as opposed to automatic, corporate control. It has been shown that occupants are adaptive to a wider range of internal temperature and more productive if able to open a window.

Thus while the automated building offers the opportunity to constantly retune to changing occupant needs, evidence suggests that this is not received in a user-friendly way. The automated building may therefore not find favour with building occupants and their managers. In this respect, arguably the prime criterion for success, this approach could be considered unsustainable.

The design stage

How a building will cope with the future will be determined in large part at the design stage. Alternative words to 'cope' could include 'accommodate', 'survive' and 'resist', each offering a rather different relationship with the future. There could be further nuances. For instance, a building that *accommodates* a new use houses it whereas a building that *accommodates to* a new use is somehow adapted to facilitate the new use. Similarly, 'survival' may be a situation of bare survival or may be a matter of continuing to thrive. Some building designs may be better able to cope with change.

The most important aspect is for the architect and other advisors and the client to ask searching questions. 'The dumbest question is the one you didn't ask' (unattributed). This searching time may also allow use of one or more of what Mark MacCormack, founder of the International Marketing Group, has called 'three hard-to-say things':

- I don't know
- I was wrong
- I need help.

Design should be informed and influenced by feedback from previous designs; how they have performed, including their ability to cope with change. It is also important to keep up with changing business and other environments, to detect and project trends and to anticipate the direction, extent and timing of change. This is a tall order and it is impossible to get everything right. Perhaps it is worth considering at a very early stage in the development of ideas, in addition to Scheme A, Scheme B, etc., not only the 'no change' option but also an 'all change' option.

Criteria for the assessment and evaluation of alternatives will also need to be developed and agreed by the parties. 'Briefing' has been recognised as an element of practice for which architects and others are not generally well educated or trained (Barrett & Stanley, 1999). At the same time, textbooks on cost planning almost routinely stress how scope for influencing the design or cost of a building declines rapidly with the passage of time beyond the feasibility and sketch design stages. This is not to say that changes cannot be made late in the design stage, or during construction, but that such changes will be disproportionately disruptive and therefore expensive. Circumstances may have changed such that a design modification is imperative; better to change than to press on and end up with a building poorly related to the newly perceived need.

Questions that may be helpful at the briefing stage could include the following.

- Who needs to be asked to gain a picture of possible futures?
- What is the desired degree of change to design for? How much can be allowed for?

- What will be the effect on the client and/or the building of not accommodating a particular change?
- Where will change impact most and where may there be opportunity to respond?
- When will effects be noticeable and what will trigger consequential changes in or to the building?
- How likely is the projected change? What else may happen in the intervening period?
- Why is the building needed and why now?

Forecasting the future

This is difficult and has provided useful projects for a variety of management consultants, centres, institutes and units, sometimes badged as 'think tanks'. Schumacher (1972) postulated a possible 'machine to foretell the future'. He distinguished (p. 220) between 'acts' and 'events' (the latter being things that 'simply happen' outside a planner's control), and between 'certainty' and 'uncertainty' in the making of forecasts, predictions and plans. The US Secretary of State for Defence, Donald Rumsfeld, has differentiated between 'known unknowns' and 'unknown unknowns' in making defence plans. Whilst one may 'expect the unexpected' it is difficult to plan for it. Toffler (1970) introduced the expression 'future shock' – 'the shattering stress and disorientation that we induce in individuals by subjecting them to too much change in too short a time'.

Living for today and planning for posterity

It could be argued that planning for a future that is so difficult to discern is likely to be so wide of the mark as to be not worthwhile. It is said that 'to fail to plan is to plan to fail' but it is truer to say that to plan for the future risks planning for the wrong future. Certainly, it seems justifiable only to plan for the known present and perhaps the 'foreseeable' future. There may also be value in having contingency or 'disaster recovery' plans for the 'worst case scenario(s)'.

Good stewardship of assets passed down by previous generations, in order to hand them on in good condition to future generations, may be seen as virtuous and honourable. Whether or not it is the best plan will depend on the criteria for assessment. Planning for the future is hard enough to determine without being further 'hamstrung' by what is bequeathed by the past. That may seem disrespectful of well-meant, perhaps sacrificial, investment built up and passed on by our forefathers; that is not intended. Future generations must be allowed to make their own dispositions as they see fit in the light of prevailing need and circumstance, unencumbered by debt or duty to an inherited obligation.

Who pays, for what, how much and when?

There was discussion in the previous chapter about placing a present value on future (and past) building-related activities. It is suggested here (and not for the first time) that, in the absence of a compelling rationale to the contrary, today's needs should be met from today's resources. By inference, if today's resources are inadequate then perhaps it is appropriate to revisit declared needs to ascertain whether on re-examination some 'needs' may be 'desires'. It may be possible to construct a 'business case' for borrowing forward. There is then a risk that a future generation may become saddled with resultant debt. Although there may be circumstances where it may be possible and desirable to be generous in expenditure, it is not appropriate to be profligate and certainly not with the hard-won results of someone else's efforts. In essence, income and outcome should be in balance.

Lifecycle costing

There has been periodic reference to changing values over time and to attempts to reconcile these. In relation, for instance, to development of the concept of 'just-in-time' maintenance (Chapter 4), there was discussion of the influences on practice of planned preventive maintenance and of lifecycle costing over the last third of the 20th century. Lifecycle costing will be discussed in the next chapter. Suffice it to say here that the concept is about trying to calculate and evaluate the cost of a building beyond its mere construction cost. This is fraught with potential problems.

Anyone who has been involved in assessing construction costs – whether as building economist, quantity surveyor, architect, estimator, director or client – will readily recognise many difficulties. Estimates are invariably based on historic data, updated in the light of experience and adjusted according to a view of future economic directions. Thus tender prices invariably differ quite substantially from one contractor to another and from the original estimate. There are also differences between the accepted tender and the final account or cost at 'completion' and handover to the client. In the wake of the Egan report increasing attention is being given to achieving greater predictability of out-turn cost.

If it is difficult to estimate construction costs for today or the near future, how much more difficult it is to estimate the cost of operating a building over its lifetime. This is where lifecycle or whole-life costing comes in. The building which is cheapest to construct may be more expensive to run and maintain than a building which was a more expensive construction and thus the costs of the 'cheaper' building may exceed those of the 'more expensive' one over time. Running costs – energy, cleaning, repairs, renewals, upgradings, etc. – over the lifetime of a building will typically be several times its initial cost. Lifecycle costing (LCC) is a way of trying to assess those costs over time. It could be used to justify construction of more expensive buildings, although Egan argued that UK buildings cost typically some 30% more than in the USA and that therefore significantly cheaper construction should be sought.

Of course, estimates of costs in the future are just that – estimates. Nor is it certain just when those costs will fall or if they will fall at all. The total cost for a building will be made up of a number of interventions and there will be a range of outcomes dependent upon decisions, whether explicit or implicit, about 'quality'. Examples could include:

- daily cleaning – perhaps hourly in popular public places, e.g. toilets at railway and bus stations, airports, motorway service stations, with periodic replenishment of paper, soap, towels, etc.
- less frequent but deeper cleaning – 'spring cleaning'
- different cleaning regimes, with consequential investments in equipment, training, energy use, 'consumables', for different surfaces, e.g. carpet, tiling, vinyl sheet, etc.
- life expectancy of carpet in whole or part with respect to anticipated and actual wear and changing fashions; possible association with dust mites and related health concerns and changing perceptions
- the next 'asbestos', which is to say, a material currently thought to be appropriate and useful that is later discovered to be unsatisfactory and hazardous to health
- redecoration in relation to 'norms', e.g. external every five years, internal every seven years. Such intervals have been determined through experience down the years but with modern paint formulations, changes in substrate, e.g. from timber to PVCu, advances in application techniques, changes in climate and weathering, etc., longer or shorter intervals may be appropriate
- changes in perception of how long visible decay, defect or deficiency may be tolerated before remedial action
- inspection and maintenance regimes such as lubrication of mechanical systems and cleaning of air filters and the relationship between these and health and safety demands, often enhanced following incidences of disease or disaster. For instance, deaths in rented domestic premises due to carbon monoxide poisoning have given rise to legislation requiring an annual inspection of gas appliances and associated certification. Periodic outbreaks of legionnaire's disease remind us of the need for continuing vigilance and documentation regimes in relation to water-cooled air handling systems.

More traditional materials and techniques tend to be tried and tested. By comparison, new materials and methods are relatively untried and their testing limited. Of course, even a much used material may be used in a novel way or in a more arduous environment. In such circumstances it may be possible to extrapolate anticipated performance and thereby maintenance and renewal intervals but special testing may be called for. Sometimes such testing and evidence of its outcomes may be demanded to demonstrate compliance with a requirement. It is common, for instance, for manufacturers to declare conformity with relevant national and international standards and this will usually involve the commissioning of tests carried out in accordance with procedures laid down in the standard itself. These tests are devised with a number of factors in mind, such as:

- practicality
- how well the test replicates or resembles the 'real-life' situation
- costs
- reconciling of perhaps passionately held varying views
- reliability, e.g. will the same, or similar, test equipment consistently produce the same results
- what degree of variability may be acceptable.

Typically, standards institutions will include representatives from various stakeholder bodies, including, for instance: manufacturers and installers and their trade bodies, designers and specifiers and related professional bodies, research institutions, universities, civil servants, people with technical and legal expertise. It is beyond the scope of this book to critique the validity or otherwise of any particular test; suffice it to say here that any test has its limitations and particularly in terms of its value in predicting future performance.

A useful examination of some of the problems and techniques used to predict the lives of building components is provided by Ashworth (1996), who states: 'Any forecast of a future event, while utilising previously recorded performance data, will always be influenced by prevailing conditions and future expectations'.

Sustainable future

The increasing attention given to sustainability since *Our Common Future* (WCED, 1989) has been referred to in previous chapters. An earlier example of professional attention to the design of buildings that would accommodate the needs of the future is the 'Long Life, Low Energy, Loose Fit' (LL/LE/LF) approach espoused over a quarter of a century ago by Alex Gordon (1974). He was then President of the Royal Institute of British Architects and wrote at an interesting time in the development of understanding of architects' involvement in and responsibility for wider 'environmental' issues. Previous RIBA presidents could be seen more as advocates for the art of Architecture with a capital A, at some remove from concerns of the 'man in the street' for a more humane, human-scaled built environment. He was followed by Rod Hackney, the 'community architect' of the housing rehabilitation scheme at Black Road, Macclesfield. This was a period of reaction to failures of the past represented by leaking flat felted roofs, rotting timber windows, high-rise tower blocks, concrete large panel systems (LPS), brought to a head by the gas explosion and consequential 'progressive collapse' of Ronan Point in east London. This effectively heralded the end of wholesale slum clearance and redevelopment programmes. The demolition rate has for some time been below 0.1% per annum; at this rate the average UK dwelling will be expected to last for more than a thousand years. This may be a desirable end although difficult to envisage perhaps when surveying the construction and condition of those homes.

Gordon's LL/LE/LF concept was born in a context of concerns about energy, specifically oil. Against a background of two wars against Israel, the 1970s saw a growing realisation in the Middle East of the value of their oil. The Organisation of Petroleum Exporting Countries (OPEC), under the guidance of Sheikh Yamani, was able to secure substantial uplifts in the price of a barrel of oil, with consequential increases in the price of petrol and petroleum-based products. This had a substantial inflationary effect on Western economies, which had developed dependencies on oil. Amongst the effects in the UK was a substantial amendment to Building Regulations, including in England and Wales the introduction of minimum standards of thermal insulation, first for housing and then for other buildings. Indeed, there was argument about the legitimacy of regulating energy use in this way, it being contended that this was not a matter of health or safety. Today, the relationship between health and the ability or inability to economically heat a building through a British winter would be more readily recognised. 'Fuel poverty' has become an issue as fuel prices have increased further.

LL/LE/LF saw the energy concern as part of a broader perspective, trying to frame a picture of a possible future. The 'energy question', combined with concerns indicated above about unattractive built forms and materials, suggested that then current ways of building may not be sustainable, although that word was not so commonly used then. Gordon's was an all-embracing concept with its focus firmly on looking at the future, its needs and how best to provide for it in a building. The emphasis on 'low energy' was largely focused upon providing a structure with a more highly insulated external envelope that would require low*er* energy to heat to the same standard temperature. In simple terms, existing technologies and techniques were 'tweaked' to achieve better performance, e.g. double glazing instead of single, the acceptance of insulation material within cavity wall construction, the development of 'thermal break' window mullions and transoms.

The 'Burolandschaft' office of the 1960s, with its extensive floors of 'open-plan' offices, was facilitated by the acceptance of PSALI (permanent supplementary artificial lighting of the interior parts of deep-plan office floors), at least by developers and their architects. A sceptic may have been less certain when seeing promotion of such arrangements by the then regional electricity boards of the UK within their new headquarters buildings. Office workers could personalise their 'workstations' by appropriate placement of potted plants. Office managers obtained the benefits of greater population densities and flexibility, being able to arrange and rearrange workgroups as they grew or shrank or changed. A good overview of the growth and changing organisation of the office is provided by McGregor & Then (1999). Provided that sufficient total space was available, this provided scope for Gordon's 'loose fit', although it is uncertain whether, with its high dependence on artificial lighting, the deep-plan building would score very well on the 'low-energy' scale.

Regarding the 'long life' dimension, as with low energy, attention seemed to be focused around tweaking to achieve long*er* life. For instance, hardwood windows would be specified in lieu of softwood (although, as noted in Chapter 8, not all hardwoods are more durable than softwoods). Those hardwood windows, and associated external doors, would commonly be stained instead of painted in the belief that redecoration cycles could be extended (although some systems provided less inhibition to moisture penetration and resistance to decay was not necessarily enhanced). Alternatively, aluminium or steel windows may be specified, perhaps with anodised or acrylic finishes, with little understanding at that time of issues related to the use of non-renewable bauxite and high 'embodied energy' through the high-temperature manufacturing processes.

An underlying expectation of the LL/LE/LF concept was that it would produce buildings that were 'light on the earth' – not necessarily light in weight (although they may be) but light in their use of resources to construct, to use and adapt. Of course, many vernacular building forms have developed over the centuries that are 'light on the earth', constructed of local materials such as wood or mud and often very appropriate to local climates, accommodating to high and low temperatures, rainfall, etc. This book does not set out to discuss vernacular architecture in more detail, but does consider the application of demonstrated attributes of sustainability as they may relate to more contemporary built forms and uses.

Gordon's approach has recently been reviewed and extended (Wood, 2000a), generating two alternative approaches to a sustainable building: the LEFT and the RIGHT building.

The LEFT building

Long Life; Low Energy; Loose Fit/Flexible; Low Technology

The LEFT building is a development of the Gordon concept, adding a reference to technology, which is described here as 'low' by comparison with the technology available today. The analysis below offers a critique of factors which may be appropriate for consideration under the component aspects.

Long Life

Highly durable materials and components should need little maintenance. However, such materials may be expensive to procure, they may be in short supply and non-renewable, e.g. stone, marble, and/or they may require a lot of energy to produce, e.g. kiln-fired brick. On the other hand, timber, perhaps almost infinitely renewable if appropriately managed, whilst liable to rot in damp conditions and vulnerable to insect attack, can have long life if installed and maintained in conditions of appropriate humidity. Preservatives, however, when used to prolong the longevity of timber, have often been found to have pernicious environmental effects, and we are witnessing something of a retreat, in bespoke architecture, from treated towards 'green' timber.

Low Energy

Few would contest the value of minimising or even eliminating the use of energy, especially if generated from non-renewable fossil fuels of coal, oil or gas or produced by nuclear power stations. Growth in availability of energy from renewable sources, such as wind, water and solar power, will reduce the pressure which was so apparent in the early 1970s when OPEC reduced supply and raised prices significantly. 'Zero-energy' buildings are already possible and there are several examples of 'autonomous buildings' which claim to be independent in terms of nett zero import/export. In the UK of late, some of the 'heat' has gone out of this area with reduction in energy prices due to competitive tariff negotiation following privatisation of formerly publicly managed operations. Consumers have been able to reduce their energy bills while actually increasing their consumption.

This is a difficult area politically. For instance, one way to reduce energy use and associated carbon emissions which contribute so much to 'global warming' and climate change is to tax according to use. A series of high-profile global 'summit meetings' have taken place (Montreal, Rio de Janeiro, Kyoto, Johannesburg) to agree protocols to secure reductions in pollution and improvements in environment and living conditions. However, the introduction of new taxes and their imposition on some sections of the community can be difficult. In the UK, for instance, the intention to extend VAT at full rate to energy bills was vehemently and successfully resisted by old-age pensioners, who are assiduous voters and therefore a powerful lobby!

Loose Fit/Flexible

This component relates to the creation of buildings within which change could be accommodated easily, for instance with little internal structure and by removal and erection of non-loadbearing partitions as and when required. The late 20th century saw much organisational change, especially in offices and in retail, and the provision of large floor-plate buildings facilitated this, often in out-of-town locations, destructive of urban form and city life. It may be that this has passed on 'savings' from the organisation as on-costs to the individual and to wider society and its agencies. For instance, if the employee or shopper then has to travel, perhaps buying a car in order to do that, there will be those additional costs.

Further, the sum of such decisions may promote a need to widen an existing road or build a new one or provide traffic-calming measures or a new underground station or line. There are also 'knock-on' effects on the quality of the urban environment and deserted city centres.

'Loose fit' may also imply an overprovision of space, above that actually required for the functions that were required at the time of design, in order to 'accommodate' anticipated growth. This may entail the provision of more structure and incur greater expenditure, as well as consumption of more resources. An alternative approach, less considered before Gordon, may be to promote and enable flexible employment practices such as flexi-time and home working.

Low Technology

This was not part of the Gordon concept; the author has added it here in that the technology of the 1970s was low-tech by comparison with today. Although communications systems were starting to expand, emphasis was still very much upon paper-based filing systems (maybe in triplicate), telephones (often shared) and hard-wired systems with mainframe computers. Air conditioning, however, was becoming more of an expectation, progressing from simpler forced ventilation systems. There was a growing need to accommodate perhaps substantial ventilation ductwork into designs and a wish to hide this, normally behind suspended ceilings. Sometimes, for instance in some hospitals and other large buildings, the volumes of air to be moved and size of plant required were such as to require the provision of separate 'service floors'. The often considerable and complex wiring to service computers demanded the additional provision of raised floors, with access designed in, either at predetermined points or, if warranted to provide flexibility for future reorganisation, with full access, through the lifting of floor tiles. These raised floors needed to be designed not only to take the normal foot-traffic of a building but also to withstand the more concentrated loads of office equipment such as mail-addressing and stuffing machines, photocopiers and trolleys to transport the ever-increasing amounts of paper.

The RIGHT building

Reuseability; Intelligence; Greenness; High Technology

By contrast with the LEFT building, the RIGHT building places less emphasis on the future and attempts to anticipate and plan for it. It can be seen as a building which is 'right for today and designed to be maintenance-free throughout its functional life' (Wood, 2000b).

Reuseability

Doors, frames and sections of partitions, for instance, may be taken down, put into store and reused later and/or elsewhere. Assemblies are based on component technology, with standardisation (albeit customised as appropriate - perhaps 'mass customisation'), dimensional co-ordination, etc.

A lot of work was done on dimensional co-ordination, both theoretical and practical, in the 1960s and 1970s, stimulated in the UK by the changeover from the imperial to metric system of measurement and a government push toward more 'industrialised building' (IB). The then Ministry of Housing and Local Government (MOHLG) Circular 1/68 *Metrication of Housebuilding* decreed that there would be a standard floor-to-floor height of 2600 mm. This facilitated the development of large panel concrete structures and standard staircase designs, but was less supportive of other internal component development such as internal partition walls, as floor-to-ceiling heights varied with floor depth, which was usually determined in relation to spans of suspended intermediate floors. A 300 × 300 mm grid was promoted for internal floor planning, with door and window openings planned as multiples of 300 mm, co-ordinating rather imperfectly with traditional 9-inch bricks and 2′3″ doors, in which a reasonable reuse market had been built up.

Since the 1970s there has been a retreat, in the domestic market at least, from the 100 mm metric modular increment to 'metricised' versions of the traditional, based on a 225 mm nominal (215 mm actual) brick dimension. In 'non-housing', for instance schools and the National Health Service, more substantial progress was made, with standardisation around panels in widths of 900 mm, 1200 mm and 1800 mm particularly favoured. The development of consortia, such as CLASP and SCOLA (referred to in Chapter 1), with their own research and development teams supported this drive. Since the large development programmes of the 1960s, 1970s and 1980s there has been a move away from the 'standard' or 'modern' appearance to something approaching a more domestic look, based around brick and tile. Internally, however, there is still strong allegiance to a metric grid with standard doors and partitioning sytems, lending themselves to potential reuse if appropriate organisational and storage arrangements can be made. Perhaps the modular approach as used in control systems of domestic equipment, that enables replacement of a malfunctioning module for repair or scrapping, may find wider application in building components.

The 'mass customisation' referred to goes beyond an individual customisation that turns a standard product into something more in line with one individual's taste or need. It is about the design and arrangement of a mutiplicity of components such that each customer chooses the particular combination of standard parts that they want, thus creating their own individual combination. This is the approach pioneered in relation to cars by Daewoo. To some extent this is what is offered by the UK private housebuilding industry, where purchasers are offered the standard 'Balmoral' or 'Blenheim' house type with their own choice of bathroom and kitchen layouts and fittings. Perhaps it could be applied to whole modular or 'volumetric' housing or hotel accommodation. The then Greater London Council's PSSHAK scheme in Stoke Newington in the 1970s may provide a model. PSSHAK (Public Support System Housing Assembly Kit) was based on the ideas of the Dutch architect Nicholas Habraken (1972) and provided a structural framework within which tenants could arrange and rearrange demountable partitions to suit their own individual and changing needs and desires.

The Egan Report (1998) gave renewed impetus to questions of industrialised building through the promotion of greater standardisation and prefabrication as routes to higher quality, fewer defects and faster and cheaper building.

Intelligent

Not necessarily automated, this implies a choice of building design and construction suited to the customer's needs, in essence a 'close fit' at all times, not incurring the costs of constant change and entirely consonant with teleworking, hotdesking and high rates of change facilitated by networking. Intelligence was discussed in Chapter 7 in relation to both care of a so-called intelligent building and the intelligent care of buildings generally, whether or not they were construed as 'intelligent buildings'.

This is about using intelligence in the sense of knowledge gained by investigation and experience, put together with wisdom to make informed and appropriate choices, thus arriving at the 'right' building. The right building will provide what the user needs, ascertained by intelligent professional questioning and follow-up studies at the briefing stage. This will include checking the customer's certainty about stated wants and needs and challenging responses where appropriate. Intelligence also implies thinking and learning. The concepts of the 'learning organisation' (Senge, 1992) and the 'learning building', one which learns from experience how best to respond to changing environments, may be relevant.

Greenness

This relates to issues of renewability and 'light' or 'lean' use of resources, in construction as well as in use. It takes on board issues of embodied energy and environmental impacts, including for instance implied transport considerations, by being 'the right building in the right place'. A number of generally low-density and low-intensity 'green' projects were referred to in Chapter 8 and included facets such as being earth sheltered, having turf roofs, recycling waste water, composting solid waste and being either low or zero energy in operation. These were often, though not always, semi-rural or suburban applications. Assessment methods have been devised, such as BREEAM, that attempt to make a composite if not yet fully comprehensive or universally agreed evaluation of a building. This is discussed further in the final chapter where a preliminary attempt is made to describe a model of building maintenance and develop the idea of a sustainable maintenance index or similar, putting this in the context of the urban environment.

A 'green' building may also be seen as one that replicates or mimics closely 'organic' or biological processes, an approach known as 'biomimetics'. This has been promoted by Sir Jonathon Porritt, of the think tank and ginger group Forum for the Future and advisor to the Prince of Wales and the UK government. Examples include the tensile structure of the spider's web and the natural ventilation system of the termite hill. Perhaps built equivalents of 'warm-blooded' and 'cold-blooded' animals could be imagined.

High Technology

This makes full use of capabilities such as not needing to travel to work, having buildings 'fine tuned' to changing weather, not providing a sterile, unchanging uniformity of internal conditions but being responsive to individual wants and needs. Maybe there could be facets of the 'cold-blooded' building here. A warm-blooded animal maintains a constant temperature while a cold-blooded animal takes on the temperature of its surroundings. The crude early environmental controls that we have become accustomed to, such as thermostats and photoelectric cells, were for a long time used to turn boilers and lights on and off and to maintain preset maximum and minimum temperatures and lighting levels. It is possible to envisage a control system that tracks changing external environments and moderates internal environments to suit individual occupants. The adaptive ability of people was also referred to in Chapter 7 and is applicable to the RIGHT building.

The development of programmable devices, the personal computer, mobile telephones and wire-less/infra-red controls enable personalised environmental parameters to be implemented. Video-conferencing, the Internet and networking have facilitated working across previous boundaries of time and space. The home (and 'homely') environment can be recreated at the office and the office recreated at home or on the train or plane. Many buildings, or spaces within them, have become multifunctional. The sophistication of control systems is such that living and working environments can be reconfigured, redesigned and re-engineered at the press of a button, thus creating the right environmental conditions in the right building, and the right parts of the building, at the right time.

Thus, the RIGHT building offers the individual, personalised, customer-focused building in use, in contrast to the 'LEFT' paradigm of working out everything in advance to accommodate the perceived needs of the masses, suitably conditioned to the 'mean'.

Alternative futures

The LEFT and RIGHT buildings are just two, contrasting, ways of conceptualising a building. The models are imperfect and not necessarily diametrically opposed to each other; it is more that they represent differing viewpoints in a wide spectrum of possibilities. They are intended to promote discussion and consideration of alternative models in relation to design and use of buildings.

Building care

This extends the concept of addressing the needs of the client and/or building user into the systems to be used within the building, its management and maintenance to deliver continuously appropriate building and internal environments. It is an 'intelligent' approach, integrated across space, time and personnel. It is 'professional' in that it is fully considered and not amateur, but that may not necessarily imply use of building 'professions'. Emphasis in building care is on assiduously seeking out and serving user needs. There is closeness to the user consistent with delivery of an excellent service. How can that best be delivered sustainably?

Sustainable care

It is one thing to design a sustainable building; it is another to maintain it and to do so sustainably. What factors will make for sustainable maintenance practices? There are matters related to the building's design and construction, the skills and operational organisation of maintenance personnel. For instance, is there value in having a dedicated maintenance team and/or continuous flow of work? The argument here is that people without an ongoing relationship with the building may be less concerned to care for it. Developments in maintenance have tended to focus on efficiency rather than effectiveness, losing sight of motivational matters. In a situation of skilled labour shortage, particularly in maintenance, can the necessary maintenance be secured if in essence the work is deskilled?

A key issue is that maintenance must be fully considered at the design stage. Data is not only in short supply, but little consideration is given to it in the building procurement process. Considerations should include:

- low wear
- modular replacements
- 'diagnostic' reporting
- standardised elements
- availability of 'spares'
- availability of 'intelligence'
- adjustability, e.g. ability to plane down or rehang a door.

Sustainable building

Will a sustainable building need more or less care in its use than otherwise? This would certainly warrant a substantial study. Much would depend on the materials chosen. Perhaps an equally pertinent factor, also worthy of further consideration, is the level at which the building is expected to perform over time. Is it acceptable to tolerate a reduced performance? If so, for how long and how would this be monitored?

A sustainable building may be less 'demanding' of maintenance, more accommodating of degraded performance, accepting of distortions, less liable to total breakdown or failure. The key factor here is that the maintenance response should be appropriate and timely, with focus on users rather than the building. Issues of comfort versus control, explored elsewhere, are very pertinent here. The more that maintenance and related interventions take account of the individual and localised needs rather than the corporate and centralised, the more sustainable will be the practices and thereby the building.

Problems of caring for the sustainable building

The principal problem is lack of 'intelligence' in its various guises – lack of data, particularly relevant data, lack of research into behaviour and durability of materials and lack of 'brainpower' applied to the maintenance field. There are also possible problems of interpretation. If unskilled, lowly skilled or inappropriately skilled personnel are expected to take responsibility for a building which incorporates materials or components that are unusual, or used in unfamiliar or untested circumstances, it may be difficult to 'understand' the building and what is happening to it.

There are also problems in anticipating how the sustainable building may perform over time. Available data will almost certainly relate to historical or test conditions, perhaps dissimilar to actual operational conditions. Forecasting future conditions and performance is difficult. Resolution of such problems is more likely with a continuing relationship from design to construction to operation, encouraging and facilitating responsible attitudes and actions.

Opportunities

The previous section has dwelt on possible problems related to the sustainable building but it is important not to lose sight of the opportunities presented, the nature and extent of which will depend on the design and materials of the building. For instance, simpler, less sophisticated, maybe even standardised construction may lead to fewer defects. Higher quality may also be possible; a more 'care-fully' designed building may result in greater attention to detail and thus fewer unforeseen difficulties in construction and maintenance. It is also more likely the building will be the 'right' one – a building properly conceived and designed to suit the context, environmental and operational, in which it will function.

The foregoing may allow lower initial costs in that there will be less of the usual redesigning that takes place during construction, because there will be better briefing and joint development of the brief between architect/designer and client/user. There should also be lower repair, maintenance and upgrading costs, as many such needs will be reduced or eliminated due to appropriate design. Indeed, it is possible to conceive that there may be no such costs if the building's life is fully consonant with its anticipated use and the building remains as built, deteriorating acceptably until it 'dies' and is demolished for recycling. This building may also facilitate empowerment of the individual user, with less or no central 'control'.

The maintenance-free building

It has become conventional to declare the impossibility of achieving a truly maintenance-free building but Wood (1999) considered it possible to approach the maintenance-free building through increasingly reducing maintenance need. The building may be high or low quality at time of construction, as long as it remains acceptable quality throughout its life. That life may be long or short, depending on the difference between as-built quality and the lowest acceptable in-use quality (the LAPeL - lowest acceptable performance level - referred to in Chapter 1) and the rate at which component parts deteriorate or become obsolescent.

A low-energy, even zero-energy, building can be conceived and exemplars have been constructed. However, increasing recognition of energy used in the construction stage (embodied energy) requires recalculation of the zero position. This may still be possible, especially if the building generates and distributes excess energy and if the materials are able to realise latent energy at the end of the useful life of the building when 'de-constructed'. A tight-fit building, constructed to meet today's needs rather than to meet a possibly never-realised anticipated future need, reduces initial costs. It may also be that specifications will be lower and less complicated technologies employed. Natural and self-finish materials will require no decoration or redecoration and it is also possible to specify materials that will need no cleaning or are self-cleaning. The maintenance-free building is a possibility.

The care-free building

This is similar to the maintenance-free building with the added component that there is no need for the building client or user to be anxious that maintenance may still be required. Features of this building will include:

- hassle free
- low dependency; user can cope with 'failure'
- little or nothing to go wrong
- if something goes wrong there is little or no adverse effect

- deterioration is 'liveable with'
- users move or rebuild when it no longer fits their needs well enough.

This concept is developed further in the following chapter.

Conclusion

The chapter has reviewed a range of approaches to the maintenance of buildings. There are a number of aspects of commonality or concern that can be drawn out.

- Maintenance is expensive.
- Maintenance is a function of 'design'.
- The 'maintenance-free building' is a possibility.

Design is an exercise comprising a multiplicity of decisions, not always overt and often underresourced, especially with regard to costs resulting from design decisions or lack of consideration of all implications. But care at the design stage will be repaid many times over, saving expense and anxiety. A care-free sustainable building is possible and this may be low- or high-tech. Use can be made of technology at the design stage and in a building's operation, with one informing the other. Sensors, with anticipatory and reactive systems, enable individually user-determined and controlled internal environments. Unacceptable deterioration and breakdown are anticipated and corrected before they occur, thus avoiding maintenance.

An old paradigm of sustainability would include:

- the ageing hippie of the late 20th century
- the autonomous house or community
- use of natural materials
- aversion to technology.

By contrast a new paradigm can be constructed, based on:

- design for today, not yesterday or tomorrow
- use of 'intelligence'
- upgrading through building replacement when required rather than recurrent repair
- provision of the 'right' building.

Much work has been undertaken on sustainability or sustainable development, yet there is still little agreement on what it is or how it may be achieved. The sustainable building is still at the developmental and experimental stage; more work is required to test built versions and report on their performance in use.

Summary

This chapter has focused on two widely divergent approaches to achieving a sustainable building. However, in life it is rarely a simple matter of choosing between two options! It is thus wholly appropriate to bring these together with the range of alternative approaches to building maintenance discussed in earlier chapters to encourage and facilitate a 'whole-life' or 'cradle-to-grave' assessment of building design, procurement and use. That is the subject of the next and final chapter.

References

Ashworth, A. (1996) Estimating the life expectancies of building components in life-cycle calculations. *Structural Survey* **14** (2), 4–8.

Barrett, P. & Stanley, S. (1999) *Better Construction Briefing*. Blackwell Science, Oxford.

Egan, J. (1998) *Rethinking Construction: Report of the Task Group to the Secretary of State for the Environment, Transport and the Regions*. DETR, London (www.rethinking.org.uk).

Gordon, A. (1974) Architects and resource conservation. *RIBA Journal* **81** (1), 9–12.

Habraken, N.J. (1972) *Supports: An Alternative to Mass Housing*. Architectural Press, London.

McGregor, W. & Then, D.S.S. (1999) *Facilities Management and the Business of Space*. Arnold, London, pp. 26–33.

Schumacher, E.F. (1972) *Small is Beautiful: A Study of Economics as if People Mattered*. Sphere Books, London, pp. 217–234.

Senge, P.M. (1992) *The Fifth Discipline: The Art and Practice of the Learning Organisation*. Century Business, London.

Toffler, A. (1970) *Future Shock*. Bodley Head, London.

Wood, B.R. (1999) *Sustainable Building Maintenance*. Proceedings of the Australasian Universities Building Education Association 24th Annual Conference and Catalyst '99, University of Western Sydney, 5–7 July, pp. 129–140.

Wood, B.R. (2000a) *Sustainable Building Care*. Proceedings of Conseil Internationale du Batiment Working Commission W70 Symposium: Providing Facilities Solutions to Business Challenges – Moving Towards Integrated Resources Management, Brisbane, 15–17 November, pp. 415–422.

Wood, B.R. (2000b) *Sustainability and the Right/Left Building*. Proceedings of Joint Symposium of Conseil Internationale du Batiment, Working Commissions W55 and W65, University of Reading, September.

World Commission on Environment and Development (WCED) (1989) *Our Common Future*. Oxford University Press, Oxford.

10 Cradle to Grave: Whole-Life Assessment and Implications for Design

The final chapter of this book reviews a range of building care options. It also reconsiders the effects upon the maintenance and use of a building of design decisions taken some time before and projects forward into implications for future design. It is recognised that many design decisions may benefit from taking a fuller consideration of later effects and that 'briefing' and design management need to be better resourced. Time spent on these activities may be crucial to obtaining the right, perhaps care-free building solution.

Introduction

The possibility of contractors offering 'cradle-to-grave' services in relation to buildings has been mooted (Wood, 1997) and discussed in relation to the procurement of building maintenance and care services (see Chapter 4). It was argued that building contractors were 'well placed' but not capitalising on the potential; indeed, they were yielding markets to operators from other fields. It is further contended that the division of separate responsibilities for design, construction and operation of a building is inefficient and fails to deliver either effective living and working environments or economy. A more holistic approach could be expected to deliver more appropriate buildings over time. Buildings consume in their lifetime several times their capital construction cost, so it is well worth taking time to consider the ongoing 'implied' costs of operating a proposed building while it is on the drawing board. This may support spending more on the construction, to achieve a more durable or more flexible building. More contentiously, perhaps, such analysis may suggest a less durable or lower specification proposal.

The development of longer term and deeper relationships between designers, builders and users, and their financial advisors, backers and bankers, provides strong incentives to investigate, and achieve, long-term benefits. Such benefits may include:

- lower costs
- higher profits

- higher quality
- improved cashflow
- greater certainty
- less hassle
- better understanding
- fewer complaints.

'Jam tomorrow', however, remains less attractive than 'jam today' so there will always be pressure to spend less today and to defer as much expenditure as possible into the unforeseeable future.

For many years, planned preventive maintenance (PPM) programmes have been the 'received wisdom' for cost-effective maintenance, particularly for large stocks of and/or corporately managed buildings. The principal drivers for PPM have been:

- breakdowns are at least inconvenient and perhaps unacceptable
- maintenance response is slow and unreliable
- 'economies of scale' – unit rates for long runs of similar work are cheaper than for 'one-offs' and such work is of higher quality.

However, there is little published evidence of the alleged cost-effectiveness. Indeed, the 'preventive' aspect of the approach implies that components are replaced while they still have useful residual life in them, which does not encourage sustainability.

Cradle-to-grave care

The term 'building care' is intended to convey a concern for the building and its users in a way not necessarily associated with maintenance or facilities management; it is people, not product or process focused (Wood, 2000). Maintenance has been much focused on the technical assessment of condition and rectification of defects (BRE, 1983; NBA, 1987; PSA, 1987; Richardson, 2000) and the production of large PPM programmes (Finch, 1988; Then, 1995; Jones & Collis, 1996; Pitt, 1997; Triantaphyllou *et al.*, 1997; Tsang, 1995). However, Atkinson (1998) identified that 'managerial influences underlie many errors leading to defects ... [and]... As a consequence, the continual emphasis placed by technical publications on correct technical solutions... is misplaced'. Chapman & Beck (1998) 'highlighted many problems with Stock Condition Surveys and the professionals who have undertaken them', calling into question much of the basis of the PPM programmes based upon them.

It is worth recalling here the 'standard' definition of maintenance as: 'the combination of all technical and administrative actions, including supervision actions, intended to retain an item in, or restore it to, a state in which it can perform a required function' (BS 4778, Part 3, Section 3.2: 1991). By its inclusion of 'administrative' and 'supervision' actions, this is an advance on the more commonly quoted version from BS 3811: 1984: 'a combination of any actions carried out to retain an item in, or restore it to an acceptable condition' (Chanter & Swallow, 1996; Wordsworth, 2001).

Building care has a 'softer' connotation than maintenance. There is less focus on condition and its assessment, and more on how satisfied the user is and how he or she may be satisfied better. Customer satisfaction, customer care, even delight may now be sought. The 'think big' approach of large PPM programmes, where individual user needs are subordinate to the demands of the programme, is no longer acceptable. Consideration of the customer transcends condition. This requires a continuous (or at least continual, i.e. periodic) assessment of user needs and, more specifically, the gap(s) between what is delivered and the real wants or needs. This demands a quality, scale and speed of response unlikely to be deliverable through a PPM programme as commonly configured today.

To work this way, buildings and facilities managers will need to be able to adapt and change themselves; they will need to deploy a mixture of quick and mature, informed thinking, keeping the 'big picture' in mind, combined with an ability to get alongside people. This is quite a change from the skill-set appropriate to the preparation of PPM programmes, with rational plans based upon evidence, fully considered in calm and quiet, and implemented to determined deadlines. It may be that different ways of working, with more emphasis on interpersonal and intuitive skills, may open new opportunities to groups currently underrepresented in building maintenance and management.

Of course, such 'openness' and flexibility may feel quite uncomfortable for the 'planners' among us. PPM enables firm budgets to be laid down for many years ahead and for actual and projected expenditure to be regularly and rigorously compared so that 'progress' and 'success' can be plotted and measured. Comfort can certainly be derived from having a substantial budget and a plan for its disbursement. It may be too radical to suggest a wholesale abandonment of PPM in favour of the uncertainty of dealing with problems and requests as they arise. However, it need not be a matter of deciding for one and against the other but perhaps of making a balance, which may shift over time.

Perhaps another significant shift could be achieved if spending on maintenance was seen not as expenditure on buildings but as investment in people through attending to what they want. Political drivers and providers of finance such as the banks and buildings insurance companies may be able to help in the same way that the UK motor repair industry was improved by demanding higher standards of performance.

Lifecycle costing

Over recent years a number of building economists and quantity surveyors concentrated on the development of lifecycle costing (LCC). In the *CIOB Handbook of Facilities Management*, Spedding (1994) makes reference to several protagonists. He states, for instance, that 'Brandon identified a cyclical interest in techniques which view the total cost of an asset over its lifetime as opposed to the cost of its initial provision, and that with each cycle there has been a change of name and a slight change of emphasis'. He cites Stone (1967), Southwell (1967), RICS (1986), RIBA (1985) and the Society of Chief Quantity Surveyors (1984), who have produced publications on lifecycle appraisal or lifecycle costing.

More recent work has tended to focus on what Ferry & Brandon (1991) have called the 'disadvantages' of LCC assessment, 'which explains why this technique has been seen more often in the examination room than in real life'. Flanagan *et al.* (1989) refer to the belief that 'any calculations that involve some degree of mathematical manipulation must be accurate' and Ashworth (1994) describes LCC as 'a combination of calculation and judgement'. Chanter & Swallow (1996) and Wordsworth (2001) give comprehensive coverage of LCC and financial appraisal techniques so little need be said here.

The 'scientific' approach suggested by LCC has, however, as with PPM, often failed in practice because sufficient funds were not available, often being cut or deferred in budget decisions, because maintenance is not a high-profile activity and because maintenance managers also have been lowly valued. It is also possible to argue that the calculations for LCC are so dependent upon the assumptions made for inflation, interest rates, decay and depreciation that it is possible to so select such rates as to support almost any desired 'result'.

Economic factors

There is always a tendency when considering a building project to focus on the estimated tender price for the construction as the main cost to the client. That is to ignore the significant costs involved in engaging the design team, the fitting out and moving in costs and all the costs of running the building and the operations therein, including the costs of repairing and upgrading the building from time to time.

In LCC and its variants, there is an expectation of being able to calculate a 'right' or 'best' answer or solution to the 'problem'. Much depends on what was included in the calculation, the assumed life expectancy of components and the interest rate selected. Similar criticisms can be levelled at attempts to assess embodied energy and to create some kind of a 'sustainability index'. Estimates have to be made of how comparatively limited or renewable are resources. Such estimations have limitations. How easy to forecast, and to what degrees of accuracy, are depletions and deterioration?

What is 'life'?

The life of the building is often little considered at the briefing stage or glib assumptions are made. For instance, it was common in the postwar years to assume a design life of 60 years for public sector housing, that being the period set for repayment of finance from the Public Works Loan Board. However, many tower blocks and high-density housing designed in the 1960s failed to make it even halfway because of social and structural problems. Would it be appropriate to plan that after 60 years of life a building should be demolished and a new one built? Or might the building be due at that time for a major refurbishment? Needleman (1965) argued that housing rehabilitation schemes were cheaper than demolition and rebuilding and the UK government deemed that such schemes should have a design life of 30 years, half that of new build. The rationales for these 'life' requirements are unclear. However, there is perhaps a case for suggesting that if the 60-year and 30-year figures represent the dates at which the building costs (including interest payments) have at last been repaid, then the building should last some time beyond, free of debt.

How could a designer be sure that a building could be expected to last 60 years? The first national Building Regulations (1965) had controls over 'short life' and deleterious materials, now encapsulated within a wider 'fitness for purpose' requirement, but firm figures for life expectancy of a particular material or component were, and continue to be, hard to find. NBA Construction Consultants (1985) produced a set of data of limited value assembled from various sources. It is difficult to plan on the basis of a quoted life expectancy for a slate roof covering of between 60 and 600 years, although it could perhaps be held to be sufficiently 'authoritative' to support a 'design life' of 60 years.

The difficulty of assessing the durability of materials and components in 'real life' or simulated laboratory conditions has been discussed in previous chapters. It is also difficult to find or develop consensus. For instance, for assessing the durability of a brick, BS 3921: 1985 recognised that 'frost resistance' should be a criterion. In the UK, continual 'freeze–thaw' actions through the colder months can be destructive of the integrity of the brick, especially its face, causing sometimes significant spalling. It has been difficult, however, to gain unanimity about how to measure the freeze–thaw cycle. Is it about degrees below freezing; duration of freezing; rapidity of freezing and/or thawing; number of cycles; shape of mortar joint; exposure category; wind-chill; other factors?

Some aspects may be comparatively easier to estimate, for instance where data has been built up over a long period, and conventions developed for its use. Examples of such databases could be:

- the Building Cost Information Service (BCIS) of the RICS for new build
- the Building Maintenance Index (BMI) Building Maintenance Price Book
- 'Spon's' and 'Wessex'
- National Building Specification (NBS)
- National Schedules of Rates (NSR)
- the Standard Method of Measurement (SMM) and
- the CIOB Code of Estimating Practice.

Estimation of the additional costs of providing a building with interchangeable or more durable components, or with larger floor spans or more space, is a straightforward task for a quantity surveyor. However, it is harder to estimate reliably the life expectation of a component. How is 'life' to be defined and determined?

The 'life' of a domestic kitchen, for instance, is more likely to be determined by changing technology or fashion than by physical determinants of decay or decomposition. A single electrical socket may have been adequate in the 1940s or 1950s but is wholly inadequate for today's gamut of kitchen equipment. How much of today's technology was known in 1940? Could its penetration have been accurately forecast? Similarly with computing and information technology in offices, changing shopping patterns, sophisticated sensing, security and control systems, to name just a few. Indeed, the 'office' itself was an innovation. The Uffizi in Firenze (Florence), built between 1560 and 1574, is 'probably the first purpose-built office' (McGregor & Then, 1999). 'Uffizi' means offices. The demise of the office is often forecast with the anticipated growth of 'tele-cottaging' and return to 'homeworking'.

Anticipating possible futures is very difficult and uncertain. Costs could be incurred in providing for anticipated futures that do not come to fruition. For instance, in the 1970s and 1980s it was common to provide office buildings with deep service zones above suspended ceilings and/or beneath raised access floors to allow for not only the wiring and air-conditioning ducts then needed but also for all the extra services certain to be required in the future. Today, much of the wiring can be replaced by wire-less communication and air conditioning is not favoured as it once was. The flexibility afforded by those service zones was provided at a cost for an unrealised benefit.

Design for low or no maintenance

In 1990, the RICS revised and reissued its Practice Note No. 4 on Building Planned Maintenance. The institution argued that:

> '... the design and maintenance processes in the construction industry need to be more closely allied as in the motor industry where design and subsequent maintenance frequently have an equal consideration. Early discussion between the design and maintenance organisations can result in a building with lower maintenance liability at a marginally increased initial cost, the economics of which can be easily demonstrated.'

Certainly, if maintenance costs are to be controlled there are implications that must be considered at the design stage. This must include not only the durability and maintenance demand of materials and components, but also consideration of how the building is to be used, and potentially abused, and whether and how change is to be accommodated.

Buildings are not used as they once were. Offices and retail premises previously staffed largely from 9 a.m. to 5 p.m. are increasingly open all day, every day. There is no 'downtime' in which maintenance work can be carried out. Users expect their working and leisure environments to function effectively at all times; when something goes wrong they want it fixed straight away, not to have to fit in with some predetermined programme.

There is also more concern and care for the stewardship of 'Planet Earth' and the use of finite resources. It is not acceptable to consign to landfill sites thousands of fluorescent tubes replaced while they still have many hours of life left in them. Attention needs to be given at the design stage to the anticipated life of components and of whole buildings. The possibility of a 'no maintenance' building, run from birth to death without intervention, was postulated in Chapter 9.

Design and sustainability

How a building is to be maintained and how long it may last are very much influenced, even if not specifically determined, at the design stage. However, briefing is often given scant attention. The RIBA issued advice within the Plan of Work (1964); Stages A and B: Inception and Feasibility would be informed by Stage M: Feedback from similar and recently completely projects. Unfortunately, few building clients are repeat commissioners of construction, so learning from previous projects is difficult for them. Arguably, in any case, the newly completed building represents yesterday's thoughts on what was required rather than today's thoughts or tomorrow's. Architects' fees for Stages A and B have historically been chargeable on an hourly-rate basis and not seen as part of the standard fee for the design of the building and the supervision of its construction. For these, a fee based on a percentage of the construction cost has long been the norm, based on the RIBA's once mandatory, now recommended, fee scale. This has tended to constrain expenditure of time and money on the briefing activity and consequently to cut out opportunity to consider alternative approaches, and this is detrimental to the best long-term interests of clients and users. These and 'Planet Earth' are more likely to be well served if greater priority is attached to full, and frank if necessary, discussion at the earliest stage.

Who decides on the appropriate design? Few company boards have property professionals as members although some, including multinational energy 'giants' such as BP and Exxon, have developed public environmental policy documents. Local planning and building control authorities have permissive powers of control; that is to say, that they approve or reject the proposals of others, so they have limited ability to promote the development of more sustainable designs. Much depends on motivation of architects and their clients to advance the art and practice of sustainable building.

Significant determinants of the sustainability or otherwise of a building include:

- location
- materials
- maintenance
- construction
- energy use
- meeting user needs.

These matters can be taken into account to varying degrees at the design stage. For instance, in terms of location this may be predetermined – for example, a new building as part of an existing complex already owned by the client. In other instances the choice of location may be almost infinite in today's global business environment. The UK Building Research Establishment's Environmental Assessment Method (BRE, 1993), based on a points tally, allocated a number of positive points to proximity to a railway station and negative points for car-parking provision. Thus a development planned without a rail station adjacent and dependent upon access by car would be unsustainable whatever the details of its design or construction.

Defects

If a building is to be run with minimal maintenance, let alone be maintenance free, it will be helpful if it starts its life without defects. Defects occur in buildings for a range of reasons. They represent a waste of resources, time and money, a difference of expectation of what was required, whether it be a small or large defect. The *Oxford English Dictionary* (1989) defines a defect as 'the fact of being wanting or falling short; lack or absence of something essential to completeness; deficiency; a shortcoming or failing; a fault, blemish, flaw, imperfection...'. The Building Research Establishment (1982) defined a building defect as 'a shortfall in performance as a result of a building fault' (defined as 'a departure from good practice as defined by criteria in Building Regulations, British Standards and Codes, the published recommendations of recognised authoritative bodies'. BS EN ISO 8402:1995 defined a defect as 'nonfulfilment of an intended usage requirement or reasonable expectation, including one concerned with safety'.

Defects are a matter of concern for a range of reasons.

- The client is entitled to what he or she is paying for; a defect represents a shortfall on that.
- Defects take time to rectify and this may delay completion and handover of the building.
- Much time and money may also be expended in identifying possible causes of defects, especially to attach blame.
- There may be significant disruption and consequential loss if defects are to be corrected in occupied buildings.

- Resources spent on defect rectification are not available for use elsewhere.
- Defects represent inefficiencies in construction procurement processes.

In the UK reports commissioned by government and august bodies over the last 60 years or so have consistently criticised the construction industry over its performance. In the decades following the Second World War the context was one of slum clearance and rebuilding. Huge numbers of new dwellings were required and the emphasis was on organisation and mobilisation of the industry to meet that challenge. A succession of reports were referred to in Chapter 2 (Simon, 1944; Emmerson, 1962; Banwell, 1964; NEDO, 1983). More recently, attention has focused more on qualitative issues, identifying and satisfying (perhaps surpassing) client needs (Latham, 1994; Construction Round Table, 1995; Egan, 1998). Egan has estimated that in the UK over £1 billion was spent annually in rectifying defective construction.

A number of 'post-Egan' initiatives have been instigated including for instance, the Construction Best Practice Programme (CBPP), Construction Productivity Network, Housing Quality Forum and Movement for Innovation (M4I). Benchmarking clubs have been set up and key performance indicators (KPIs) published (www.dti.gov/construction/kpi). One of the ten KPIs relates to defects, with facets including:

- identifying the relevant areas of performance, e.g. time taken; delay to 'date for completion'; cost; closeness to estimated cost; issues of quality; ease of dispute resolution; and so on
- determining appropriate items to measure and methods of measurement
- who will measure, when, etc., including matters of confidentiality, 'spying', etc.
- standards already achieved and those to be striven for
- realistic timescales for improvement.

Toward 'zero defects'

Egan, informed by the Construction Round Table, postulated a reduction in defects of 20% per annum over a five-year period. How could that be achieved or even attempted? A classic management approach would be to identify the items on which improvement might be concentrated to greatest and soonest effect. That involves recognising not only the most expensive and extensive defects but those most susceptible of extermination. The author was involved in the BRE (1982) study referred to in Chapter 4, which was aimed at reducing defects in new housing construction in the UK. It was decided to publish concise checklists based on defects actually found on a number of sites inspected. The study identified that nearly half of defects stemmed from matters of design or specification rather than construction.

The BRE subsequently produced its series of *Defect Action Sheets* (1982–90) and more recently *Good Building Guides*. However, Atkinson (1998), as stated earlier, established that defects arise not so much from lack of technical information but more through matters of management and motivation. The fragmentation and adversarial nature of the construction industry deliberated upon by Latham are at the core of the problem. This is exacerbated by skills shortages and low levels of education and training, within a 'culture' in the industry that has tolerated, and in many ways encouraged, shoddy workmanship. The construction industry has what Porter (1985) described as 'low barriers to entry', although a number of initiatives have been taken in recent years (Wood, 1998), including the Construction Skills Certification Scheme (CSCS) championed by Tony Merricks of Balfour Beatty.

The Major Contractors Group has set an objective of an 'all qualified workforce' for 2003 and the Construction Industry Training Board, now under the chairmanship of Sir Michael Latham, is giving added impetus to wider development of skills and promotion of the CSCS. Although the CSCS was developed initially as an aid to householders to combat the 'cowboy', it could provide incentive for building operatives and employers to demand upskilling and thus have a major impact on improving the image of the industry and its output. Through further, higher and continuing education and training, major reductions in the incidence of defects could be expected. The achievement of the zero-defect building is a possibility.

Quest for quality

The years since the Second World War have seen much attention to construction and procurement methods aimed at improving the economy and efficiency of the industry. In the immediate postwar years the then Ministry of Works instigated a series of building studies supported by the work of the then Building Research Station. Efforts were focused on trying to understand the physics and performance of building materials and how to improve the economy of building construction. Shortages of traditional building materials gave impetus to new ways of working and use of less conventional materials. This gave rise, for instance, to the development of building 'systems', based on precast concrete panels and steel or aluminium frames – the birth of the 'prefab'. Replacing homes lost through bombing and slum clearance were the main drivers of technological advance in domestic construction, reaching a climax in the 'industrialised building' of the 1960s. This came to an abrupt end with the gas explosion and progressive collapse of the Ronan Point tower block in Newham, London, in 1968, referred to in Chapter 1. Since then, domestic construction has followed a fairly conservative path using largely traditional construction techniques based around brick cavity walls and tiled roofs.

By comparison, the commercial sector has seen substantial advances, especially in building tall and providing highly controlled office environments. Technological advances have included air conditioning and associated suspended ceiling systems, access floors, intelligent glazing and sophisticated environmental control systems. It remains to be seen whether the disaster of 11 September 2001, with the destruction of the New York World Trade Centre, will refocus future developments.

Some attention has been given to quality relating to issues such as specifications, whole-life costs and sustainability. For some, quality may be about using more expensive materials, for instance a marble-finished foyer in lieu of plastered blockwork; others may be more concerned about increasing durability or specifying products from renewable resources. The suggestion of Elkin *et al.* (1991) that 'the most effective mechanism for reducing many of the environmental impacts associated with building materials is to design for durability' was discussed in Chapter 8. Davidson & MacEwen's argument (1983) for 'long-life, high-quality buildings that require little maintenance and are renewed at long intervals rather than low-quality structures that are replaced more frequently' was discussed in Chapter 9 in relation to LEFT and RIGHT buildings. Durability would not be everyone's measure of 'quality'.

Achieving quality

How is quality achieved? For many the focus may be upon quality *control,* analogous to inspection at the end of a production line to reject finished products that don't meet the specified standard, so that only good product is despatched. In the construction context this has produced a situation in which although attention may be paid to 'standards' at the design stage, there may be little attention to quality during construction, relying upon a Clerk of Works (if there is one) to pick up defects. This reinforces the adversarial relationships identified by Latham, bolstering the cowboy culture in which operatives, and their managers, expect to 'get away with' non-conforming work. Excuses given for poor work include:

- the architect's drawing was unclear or impossible to build or late
- the preceding work, by others, was inadequate or inaccurate in some way
- to correct it now will cause delay and/or will always show where it was rectified
- no-one noticed or complained earlier
- 'that's how I've always done it'
- 'what do you expect for that price?'

Recognising that attempts to control quality 'after the event' are relatively ineffective and expensive, the 1980s and 1990s saw a switch of focus 'upstream'. Cole (1999), referred to in Chapter 3, quoted a CEO who stated: 'My epiphany was when I realised the significance of the simple but profound observation that it is always cheaper to do things once than twice'.

Emphasis has shifted to trying to assure quality through systems. ISO 9000 certification, however, does not measure the quality of a product or service, it only confirms that there are fully documented quality control procedures and that they are being adhered to. For those who have followed the ISO 9000 route, there has been a tendency to create a mountain of paper and a 'tick box' mentality. This continues to detract from recognition that quality is everyone's responsibility.

Avoiding defects
The proposition presented here is that of the CEO quoted above, that it is cheaper to do things once than to have to redo or correct them. It is also more rewarding personally in terms of 'a job well done', more likely that the project will be completed on time and that the client and building users will be satisfied. A satisfied client is also more likely to become a repeat client. All of these are good reasons to concentrate on quality but how to achieve this highly desirable situation?

To engender an expectation that everyone will contribute to quality, its achievement and enhancement, it is important that all are fully engaged from the outset. Individuals will lose interest and enthusiasm if they see others, especially their managers, settling for less than the best. Managers must set an excellent personal example. Education and training will be key. Arguably, the newer the technology in use, the less familiar people will be with it, with greater scope for error and for those errors to lead to significant problems. This will need to be anticipated and accommodated, for instance by instruction, special training and supervision, with time and money available to get things right or to correct them if not right first time.

The culture of the 'cowboy' and cutting corners and generally low levels of qualification are unhelpful. Exhortation on the part of 'the management' is unlikely to produce significant improvement; a concerted effort is required to change attitudes as well as aptitudes. Failure to avoid defects could arise from, for instance:

- lack of knowledge and/or understanding
- lack of care or commitment
- inadequate or inappropriate materials
- incorrect instructions
- inability to carry out instructions.

The BRE study (1982) referred to earlier identified that many defects stemmed from design. It is therefore important to put effort into defect avoidance at the early stages of development. Some elements may be anticipated as more likely to give rise to defects or to defects that will be more expensive to rectify. For instance, foundations can be problematic and a rich source of defects; the value of a good site investigation is axiomatic. A study by the UK National House Building Council (NHBC, 1984) showed that in an 11-month period a total of £3.6 million was spent in rectifying shortcomings in substructure. Similarly, a flat roof may be more difficult to detail for effective rainwater disposal than a pitched roof, especially at a domestic level; the latter is also more likely to be familiar to most roofing operatives. Complex shapes, valleys, dormer windows and rooflights complicate matters, increasing risks of defects. Such issues need to be borne in mind at the conceptual and feasibility stages and followed through as detail is worked up.

Arguably the greatest contribution to defect avoidance could be made by constructors. However, they are often not involved in the process until well after the design is substantially complete. With the 'traditional' procurement route there are sharp divisions. More recent developments, such as design and build, design-build-finance-operate (DBFO), build-own-operate-transfer (BOOT), the Private Finance Initiative (PFI) and Public–Private Partnership (PPP), have endeavoured to bring contractors' experience to bear in influencing or determining design solutions. Constructors' contributions to 'buildability' are key considerations (Griffith, 1990). Although architects particularly, and others such as the Commission for Architecture and the Built Environment (CABE), express concern at loss of design quality, clients are attracted by promises of lower and more certain prices and shorter timescales. Clients also tend to prefer the prospect of 'one bum to kick' in the event of failure. Contractors' education and qualifications specifically in relation to defects and their avoidance are lacking and attention should be given to this in the construction curriculum. Learning from previous projects is also limited by the prototypical, 'one-off' nature of most designs, leading to lack of opportunities of repetition or improvement in future schemes.

The possibility of zero defects

To construct a building with no defects has long been seen as a practical impossibility, though perhaps a theoretical possibility; similarly the maintenance-free building. However, the author has postulated that this latter may now be a realisable goal and to achieve 'zero defects' is also realistic.

There is sufficient evidence to support a view that defects are both prevalent and eminently avoidable. Furthermore, information is also available to inform moves toward zero defects. What seems to be lacking is a consistent drive to improve quality or maybe even to demand and enforce quality specified in contract documents. In this regard the work done post-Egan, particularly through the best practice clubs, KPIs and benchmarking, offers real scope for developing common purpose and taking action. Applying agreed measures for monitoring and reporting also prompts regular evaluation of actions taken in the quest for continuous improvement.

It has been said that 'what gets measured gets done', so it is imperative to identify the right priorities and how to measure improvement and what will constitute success. It is important to set objectives that are SMART – strategic, measurable, achievable, realistic and time based. In the case of a large task, it may be sensible to set up a pilot project to test the validity and reasonableness of the end goal by attempting to achieve some intermediate goals or milestones. Thus, while attaining a target of zero defects on the next project may feel unachievable, to reduce defects and consequent rework by 50% over five years may seem reasonable. Progress can be reported and evaluated quarterly and annually and programmes and procedures modified accordingly.

It is important, of course, that action plans should reach back into informing and influencing designers and their designs. It has been common practice for many years for architects' drawings to contain within the title panel a box labelled 'checked by'; this is often left blank! It may be that the meaning of 'checked' is unclear – that it is no-one's designated job to check, that nobody is prepared to sign off, that there was no time allowed. The value of such a check is certainly debatable. However, a holistic overview of the project is invaluable if not essential to the attainment of zero defects. In this regard, similarity of a project to one recently undertaken is very useful. Details can be revised in the light of construction, experience and possibly of the building in use, as well as reflecting changing standards. Large clients such as the retail chain stores and commercial developers have developed long-term relationships with their supply chains, including consultants and contractors, to deliver consistent quality of buildings, just as they have with their suppliers of goods and other services. This approach has tended to produce a building that is standard in appearance. McDonalds, the fast food chain, has developed a standard layout and construction of restaurant manufactured in volumetric form in a factory that can be delivered to site and erected in a matter of days. There are also examples of other building types, for instance the housing project at Murray Grove in London's Hackney, which used similar factory-made modular units assembled on site in a fraction of the time of traditional techniques. The quality of internal finish would also be hard to achieve if done in the conditions that prevail on a normal building site.

The key constraint in achieving defect-free building is the prevailing culture of low expectation in and of the construction industry. Skill shortages, lack of understanding and care are contributory and would be improved by addressing the culture issue and associated behaviour and practices. Change can be and must be effected. For instance, the European automotive industry of the 1960s was characterised by unreliable new vehicles, poor servicing and the untrustworthy 'second-hand car dealer'. Perhaps responding to the threat from Japan, the motor industry is now typified by warranties of three years or longer with servicing at 15 000 km intervals to predetermined schedules by authorised dealers with qualified mechanics and pricing according to published 'menus'. In another field, the manufacture of silicon chips, Hewlett-Packard made great strides in quality through a joint venture with Yokogawa, referred to in Chapter 3, where defects were reduced over two years by a factor of 100. By contrast, the consultation paper on combating cowboy builders (DETR, 1998) quoted 'Latest figures from the Office of Fair Trading [1996] indicate over 93,000 complaints about home maintenance up by 4% on the previous year', very much in line with the previous ten years' average rise of 5% a year. This situation needs reversing; to satisfy clients, these figures should be declining, not rising!

Problems reach back into the supply chain. Fewer school students are studying mathematics or science. Many construction operatives are functionally illiterate or innumerate. The industry has much to do if it is to have a workforce able to build successfully with more advanced technologies.

However, change is under way in the UK. The Major Contractors Group has made a compact with the Construction Clients Forum, a group of major repeat customers that includes, for instance, Tesco, Marks and Spencer and the British Airports Authority, undertaking that by 2003 they will employ a fully qualified and CSCS card-carrying workforce. This is a substantial commitment and should give some expectation of consistently higher standards of quality and delivery. There are doubts about the ability to hit the target. Intermediate targets have not been met and concessions have been made in relation to 'grandfather rights'. Under this provision, craftspeople who can demonstrate a satisfactory period of experience will be exempted from the requirement to have a relevant qualification. There are continuing concerns about recruitment and retention of sufficient skilled personnel and needs for upskilling to meet the demands of new ways of constructing and procuring buildings.

It can be seen that there is demand for a more responsive and responsible construction industry, more professional, less adversarial, striving for greater understanding and satisfaction of client needs, and that parts of the industry at least are responding. There is much to be done, however, and the scale is such that this will take time. It is important therefore to learn from what others are doing and to share good practice. This will need trust, and the conditions for that may not yet exist (Smyth & Thompson, 1999). There remains a reticence to share experience, whether good or not. Therefore it may be more profitable to concentrate on attaining a better understanding of one's own organisation and its performance, becoming able thereby to determine reasonable standards for improvement per annum or per project, at a level that can be achieved and sustained. Bean (1997) has reported how one contractor introduced a system relating project performance to client perceptions and priorities.

Implementation may also be problematical; how to inculcate a new culture of 'zero defects'? Perhaps a consideration of how matters have changed in relation to health and safety may be instructive. There has been an attempt to secure improvement through legislation such as the Health and Safety at Work Act 1974 and the Construction Design and Management Regulations. However, while there have been impacts on the numbers of accidents, injuries and deaths, these improvements have not been sustained. More recently emphasis has been placed on employing a fully qualified workforce that receives full on- and off-site briefings related to safety. A greater commitment to training and respect for people appears to be bringing improvements.

Conclusion

Defects are expensive; they take precious time and resources to diagnose and to correct and they waste time and resources to create in the first place. £1 billion was spent in rectifying defects in UK buildings in 1997. Most defects could be avoided but this requires a good understanding of construction detailing and of what designers and clients are endeavouring to achieve. Problems occur if objectives are not shared and conflicts become difficult to resolve. Good communication is key.

Guidance to the industry hitherto has tended to focus on the production and dissemination of technical information, largely directed at architects and other designers, yet there is no evidence that this has been effective in securing improvement of construction quality. This supports the contention that defects occur more through management error or misunderstanding than through technical deficiencies. There is scope for significant reduction in the production of defects, thus improving efficiency of the construction industry and reducing the cost of building.

Although costs of rework may be difficult to quantify, the presence or absence of defects is fairly readily discernible. There may be disputes about whether or not a particular 'defect' is indeed such but arguably this arises from lack of clarity about what the client needs or wants. Clear briefs and ways of resolving uncertainties are crucial and it is therefore very helpful to have a discerning and/or experienced client working with familiar constructors and consultants. This is a route to useful learning from previous projects and injection of improvements into the next.

What tends to cut across attaining the goal of zero defects is that conflicting messages may be given or received and personal and corporate aims may diverge. Quality is often compromised by greater imperatives of time and/or cost. For instance, while zero defects may be a declared objective, damages will commonly be applied in the event of failure to complete on time. There may be rewards for completing on time and in budget but no recognition or reward for creating a defect-free building.

While it is important to reward the right things, how are those right things to be determined? It is still unusual to ask the client to identify their priorities and it can be enlightening to see not only what these are but also how they differ from those expected. This could become a significant driver of where effort should be best directed by contractors and consultants, especially if reward reflected those emphases. Perhaps in the absence of such a clear link, the most readily recognised measure of success would be the award of repeat business. Considering the disproportionately large costs of preparing estimates for unsuccessful tenders, this is a very worthwhile gain.

If 'zero defects' is more likely to be achieved in buildings with tried-and-tested technologies, this suggests a need for the testing and refinement of new technologies to encourage their use by clients, for whom otherwise conservative approaches will remain most attractive. This, rather than the technological advance itself, provides the abiding challenge.

Whilst maintenance remains 'unsexy', attention and resources have been expended on providing high-specification, high-technology, high-cost 'solutions'. For instance, extensive PPM programmes have been implemented to avoid the management 'problem' of providing an effective service that meets the needs of the customer, the building users. Many parts of buildings are overdesigned; they would continue to provide more than adequate service long after the original need for the building has passed. There is therefore much scope for revisiting the design and specification of buildings with a view to achieving less resource-intensive and more sustainable constructions.

Some organisations have recognised that their employees want more control over their working environments and that they then work more productively. Buildings, and systems within them, can be designed to enable greater personal control, for instance with personalised space, individual sensors, openable windows. Existing building users can be consulted at the briefing and design stages. Feedback can be elicited from previous projects. Maintenance systems have been developed that provide more customer-focused service.

Technological possibilities and expectations are constantly changing, arguably increasing. However, it is very difficult to predict or anticipate any particular change. It is also expensive in terms of finance and physical resource to provide for futures that may not materialise. It is therefore proposed that serious consideration should be given at the design stage to ascertaining what building is really needed, so that the right building is provided, in the right place and at the right time. That building can then be 'run to expiry' with little to no maintenance; a virtually 'care-free' building.

Summary

This chapter has brought together a number of thoughts about the relationship between design and maintenance, drawing on extensive researches and experiences of the author and others over a span of years. Design decisions are a powerful driver of maintenance needs. Together with the desires of users and budgets of building owners, they are significant determinants of maintenance demand. However, if the construction and property industries are to benefit fully from a reconsideration of that relationship between design and maintenance, then maintenance expertise and experience must be fed back routinely to inform and maybe determine design. Perhaps we could consider the potential for a holistic building life model that would encourage and facilitate such developments.

References

Ashworth, A. (1994) *Cost Studies of Buildings*. Longman, Harlow.

Atkinson, A. R. (1998) *The Role of Human Error in the Management of Construction Defects*. Proceedings of COBRA '98: RICS Construction and Building Research Conference. Oxford Brookes University, Vol. 1, pp.1–11.

Banwell, H. (1964) *Report of the Committee on the Placing and Management of Contracts for Building and Civil Engineering Work*. HMSO, London.

Bean, M. (1997) *Developing and Supporting a Trial Performance Measurement System*. Proceedings of the 2nd National Construction Marketing Conference, 3 July, Oxford Brookes University, pp. 32–36.

Building Research Establishment (1982) *Quality in Traditional Housing*. HMSO, London.

Building Research Establishment (1983) *Building Defect Action Sheets*. BRE, Garston.

BRE (1993) *BREEAM: An Environmental Assessment for New Offices*. BRE, Garston.

Chanter, B. & Swallow, P. (1996) *Building Maintenance Management*. Blackwell Science, Oxford.

Chapman, K. & Beck, M. (1998) *Recent Experiences of Housing Associations and other Registered Social Landlords in Commissioning Stock Condition Surveys*. Proceedings of COBRA '98: RICS Construction and Building Research Conference, Oxford Brookes University, Vol. 2, pp. 32–37.

Cole, R.E. (1999) *Managing Quality Fads: How American Business Learned to Play the Quality Game*. Oxford University Press, New York.

Construction Round Table (1995) *Thinking About Building*. Business Round Table, London.

Davidson, J. & MacEwen, A. (1983) *The Liveable City in the Conservation and Development Programme for the UK*. Kogan Page, London.

Department of the Environment, Transport and the Regions (1998) *Combating Cowboy Builders: A Consultation Paper*. DETR, London.

Egan, J. (1998) *Rethinking Construction*. DETR, London.

Elkin, T., McLaren, D. & Hillman, M. (1991) *Reviving the City: Towards Sustainable Urban Development*. Friends of the Earth, London.

Emmerson, H. (1962) *Survey of Problems Before the Construction Industry*. HMSO, London.

Ferry D.J. & Brandon P.S. (1991) *Cost Planning of Buildings*. Blackwell, Oxford.

Finch, E. (1988) *The Use of Multi-attribute Utility Theory in Facilities Management*. Proceedings of Conseil Internationale du Batiment Working Commission W70, Whole Life Asset Management, Heriot-Watt University, Edinburgh.

Flanagan, R., Norman, G., Meadows, J. & Robinson, G. (1989) *Life Cycle Costing*. Blackwell, Oxford.

Griffith, A. (1990) *Quality Assurance in Building*. Macmillan, London.

Jones, K. & Collis, S. (1996) Computerised maintenance management systems. *Facilities* **14** (4), 33–36.

Latham, M. (1994) *Constructing the Team*. HMSO, London.

McGregor, W. & Then, D.S.S. (1999) *Facilities Management and the Business of Space*. Arnold, London.

National Building Agency (NBA) (1987) *Common Building Defects: Diagnosis and Remedy*. Longman, Harlow.

National Economic Development Office (NEDO) (1983) *Faster Building for Industry*. HMSO, London.

National House Building Council (NHBC) (1984) *Good Housebuilding No. 6*. NHBC, Amersham.

NBA Construction Consultants (1985) *Maintenance Cycles and Life Expectancies of Building Components and Materials: A Guide to Data and Sources*. NBA-CC, London.

Needleman, L. (1965) *The Economics of Housing*. Staples Press, London.

Pitt, T.J. (1997) Data requirements for the prioritisation of building maintenance. *Facilities* **15** (3/4), 97–104.

Porter, M. (1985) *Competitive Advantage: Creating and Sustaining Superior Performance*. Collier Macmillan, London.

Property Services Agency (1987) *Common Defects in Buildings*. HMSO, London.

Richardson, B.A. (2000) *Defects and Deterioration in Buildings*, 2nd edn. E. & F.N. Spon, London.

Royal Institute of British Architects (RIBA) (1964) *Handbook.* RIBA, London.
Royal Institute of British Architects (RIBA) (1985) *Life Cycle Costs for Architects: A Draft Design Manual.* RIBA, College of Estate Management, Reading.
Royal Institution of Chartered Surveyors (RICS) (1986) *A Guide to Life Cycle Costing for Construction.* RICS, London.
Royal Institution of Chartered Surveyors (RICS) (1990) *Planned Building Maintenance: A Guidance Note.* RICS, London.
Simon, E. (1944) *The Placing and Management of Building Contracts: Report of the Central Council for Works and Buildings.* HMSO, London.
Smyth, H.J. & Thompson, N.J. (1999) *Partnering and Conditions of Trust.* Proceedings of Conseil Internationale du Batiment Joint Symposium of Working Commissions W55 and W65, University of Cape Town, September, Vol. 1, pp. 424–435.
Society of Chief Quantity Surveyors (1984) *Life Cycle Cost Planning.* SCQS, Aylesbury.
Southwell, J. (1967) *Total Building Cost Appraisal.* RICS, London.
Spedding, A. (ed) (1994) *CIOB Handbook of Facilities Management.* Longman, Harlow.
Stone, P.A. (1967) *Building Design Evaluation – Costs – Use.* E. & F.N. Spon, London.
Then, D.S.S. (1995) Computer-aided building condition survey. *Facilities* **13** (7), 23–27.
Triantaphyllou, E., Kovalerchuk, B., Mann, L. & Knapp, G.M. (1997) Determining the most important criteria in maintenance decision making. *Journal of Quality in Maintenance Engineering* **3** (1), 16–28.
Tsang, A.H.C. (1995) Condition-based maintenance: tools and decision making. *Journal of Quality in Maintenance Engineering* **1** (3), 3–17.
Wood, B.R. (1997) *Building Maintenance Service Procurement: Just in Time Maintenance.* Proceedings of Conseil Internationale du Batiment Working Commission W92 Symposium: Procurement – A Key to Innovation, University of Montreal, 20–23 May, pp. 801–811.
Wood, B.R. (1998) *Maintenance Service Development.* Proceedings of COBRA '98: RICS Construction and Building Conference, Oxford Brookes University, Vol. 2, pp. 169–177.
Wood, B.R. (2000) *Sustainability and the Right/Left Building. Proceedings of Conseil Internationale du Batiment Joint Symposium W/55/W65,* University of Reading.
Wordsworth, P. (2001) *Lee's Building Maintenance Management,* 4th edn. Blackwell Science, Oxford.

Index